# SECRETS
# OF THE
# HUMAN
# BODY

# SECRETS OF THE HUMAN BODY

CHRIS VAN TULLEKEN, XAND VAN TULLEKEN
& ANDREW COHEN

FIREFLY BOOKS

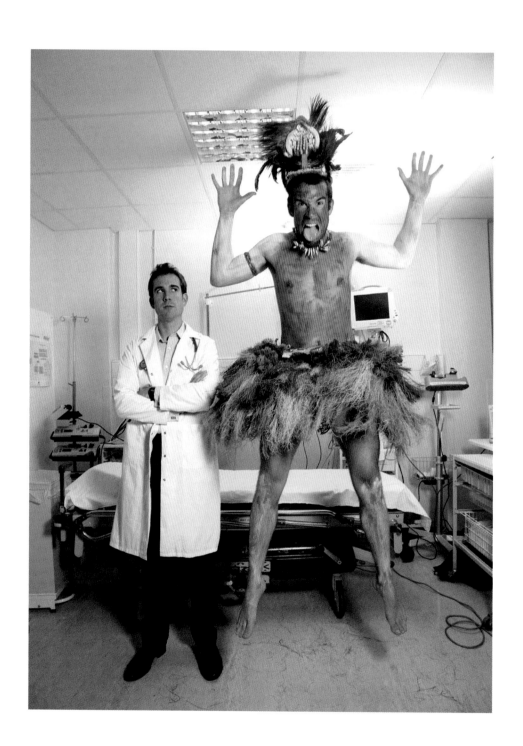

SECRETS

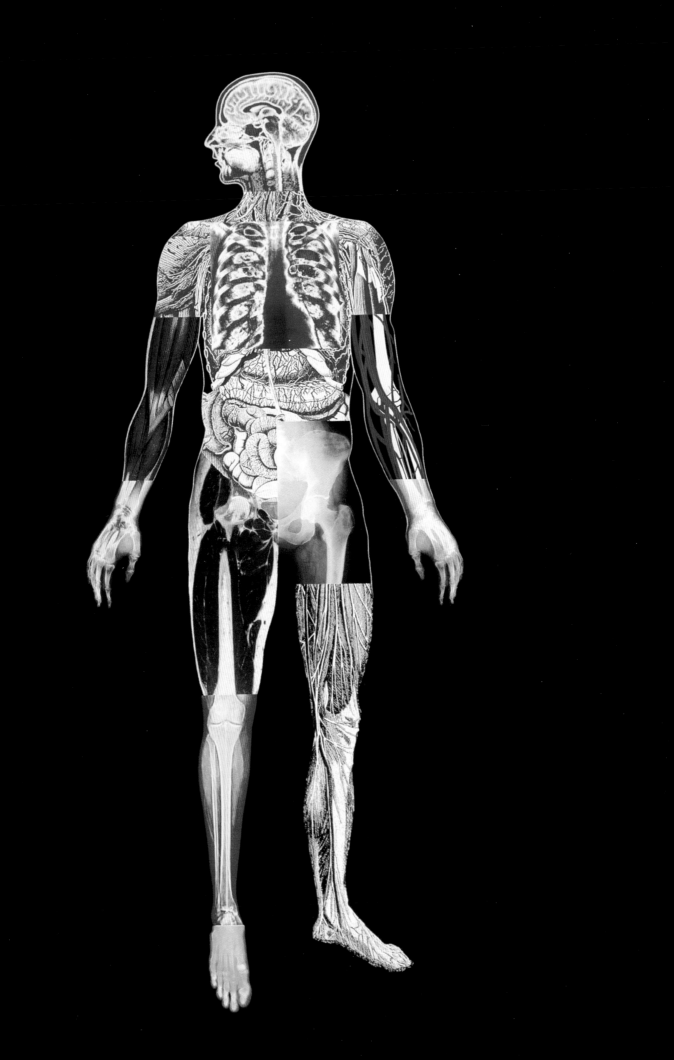

# SECRETS

The human body is in the business of keeping secrets. This was never more apparent to me than in the dissection room in my first year at medical school. The cohort of students was divided up into groups of six and we each had our own cadaver, the euphemism used for dead human body. Looking back, there was a great deal of secrecy about the whole endeavour: we were not allowed to know our bodies' names or their life story (though peculiarly we were later allowed to attend their funeral if we wished). We were not allowed to take photographs. Only medical students and staff were allowed into the room, no backstage tours for curious friends. We were meant to be respectful, which generally we were, although the atmosphere was cheerful. We were not really exploring uncharted territory, after all there isn't much anatomy left to be discovered, at least not any that I was likely to find with my scalpel and forceps. We were more like a little gang of tourists trying to get to know a new town with the help of a guide book and tour guide, in the form of our instructor Professor Hall-Craggs. And yet we were uncovering secrets: we had views of our cadavers that few people had ever seen before. How many people in your lifetime will see you stark naked? Ten? Maybe twenty? OK, maybe more depending on your job and physique, but most

**Previous spread:** Coloured scanning electron micrograph of the surface of the gastric mucosa, stomach lining, showing gastric pits. These pits are lined with the various cells that produce gastric secretions including mucus, acid, enzymes and intrinsic factor, which enables us to absorb vitamin B12.

**Opposite page:** Composite image of several methods used to visualise the anatomy of the human body. The techniques seen here are magnetic resonance imaging (MRI), X-rays (radiography), dissection and artworks.

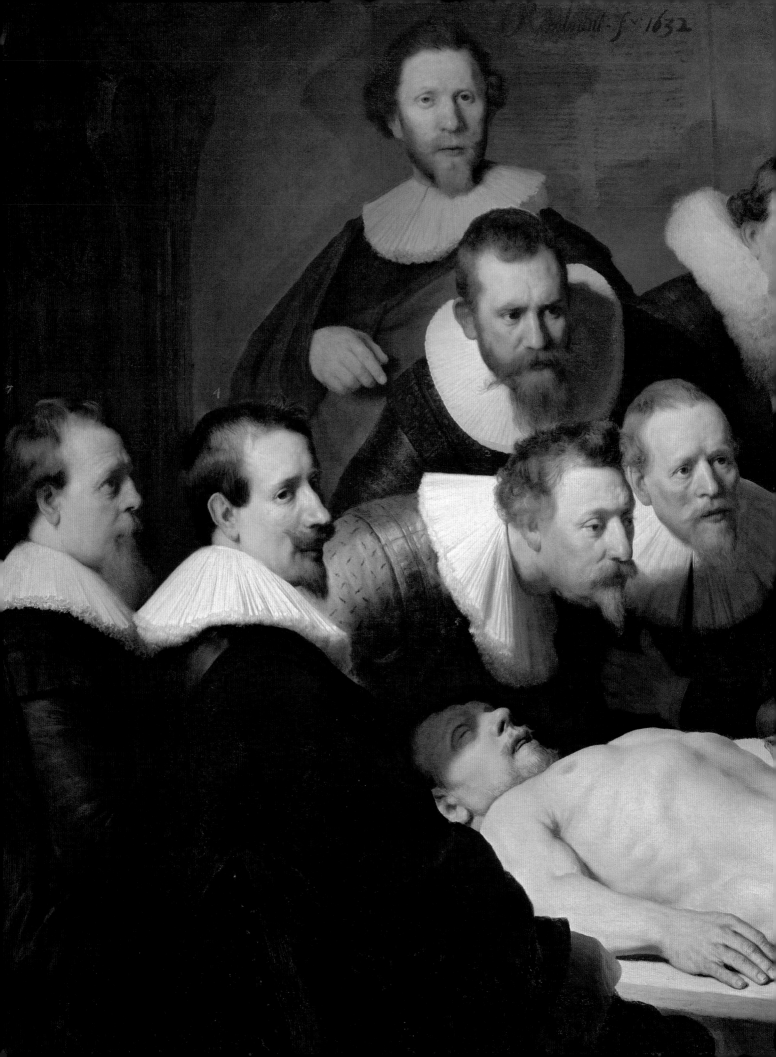

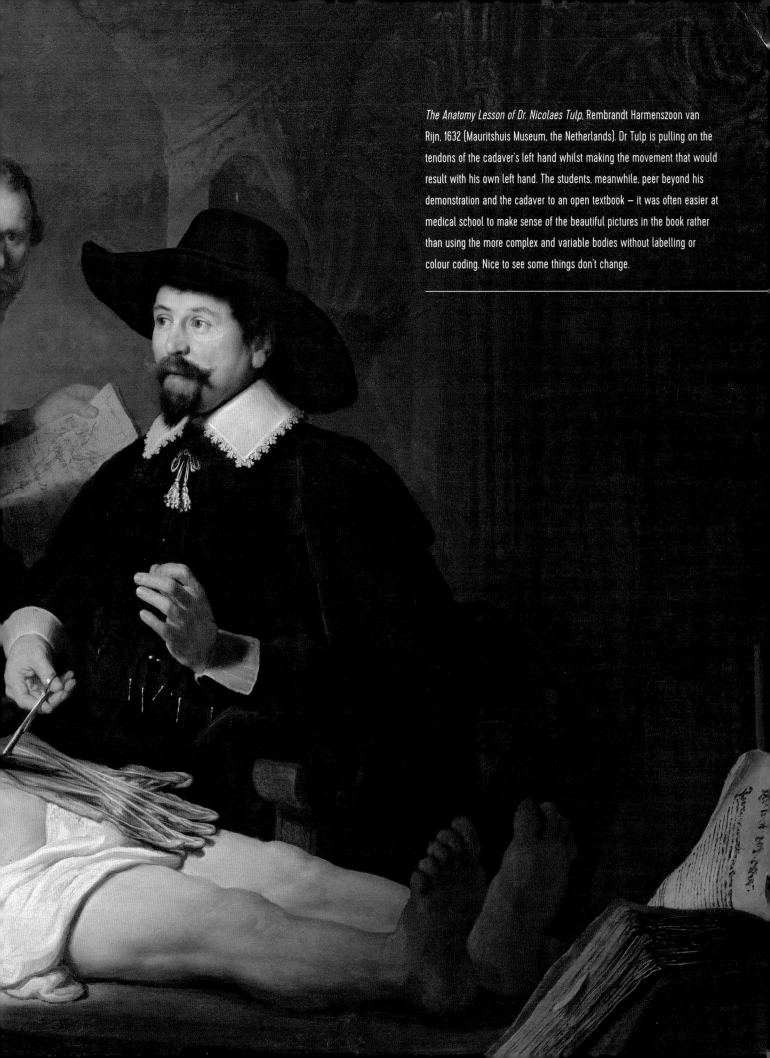

*The Anatomy Lesson of Dr. Nicolaes Tulp*, Rembrandt Harmenszoon van Rijn, 1632 (Mauritshuis Museum, the Netherlands). Dr Tulp is pulling on the tendons of the cadaver's left hand whilst making the movement that would result with his own left hand. The students, meanwhile, peer beyond his demonstration and the cadaver to an open textbook – it was often easier at medical school to make sense of the beautiful pictures in the book rather than using the more complex and variable bodies without labelling or colour coding. Nice to see some things don't change.

of us keep most of our physical bodies concealed most of the time. For many of us it was our first experience with the intimacy with which we would later have to examine living bodies (I can really only speak for myself, not my classmates, but I had seen very few naked people at age 18). And how many people have seen the inside of your body, with its various anatomical irregularities? Probably none? Perhaps a surgeon? When cutting open a body, whether alive or dead, you get a sense of seeing into a place that no one has seen before, of being privy to a secret.

My group found that our cadaver's false teeth had been left in and that they had a name engraved on them, I suppose to avoid mix-ups, which must be a concern when surrounded by older contemporaries later in life. It was a shock to think what else we didn't know about this previously anonymous man. Likewise finding his tattoos. Reminders of the secrets that were kept from us.

We got used to the smell (of chemicals, particularly formalin – not rot or decay) and to the cold slippery texture of the bodies like kelp on a beach in winter. We also got used to the squeamishness that I think we all felt with the gristle and stomach contents and vast quantities of congealed fat. I remember Professor Hall-Craggs running his hands through his hair in exasperation at my continued failure to identify the various parts of the brachial plexus. He had been dissecting all morning and his hands were covered in bits of human, and so a large globule of human fat lodged in his hair, along with a decent amount of formalin and other small bits and pieces. I was amazed and impressed by his comfort with the material of these dead people who had so generously let us cut them up. I thought that my reluctance to really get stuck in and to leave the room at the end of each session with little parts of human in my hair probably accounted for much of my ignorance. I am sure I was right. You can't be squeamish and be a good student of the human body.

We got used to the strange sight of naked humans among clothed ones day after day. Medicine involves asking people to take their clothes off, to expose their bodies at different points. We do our best to preserve patients' modesty but if you want to know what's going on with a person's body, you need to see it. We were taught two rules of thumb. If you want to examine someone's abdomen (tummy), you had better expose them 'nipples to knees' lest you miss the chest or thigh manifestations of some abdominal pathology. The other rule was 'if you don't put your finger in it you'll put your foot in it'. Meaning that we were not to shy away from rectal exams. It is frequently difficult, undignified and uncomfortable to reveal the secrets of any particular person's body. The early anatomy classes with our cadavers did a great deal to get us all used to the intimacy needed to examine and diagnose a person.

What I never got used to was the vast complexity of human anatomy. I passed my exams like everyone else. I even went on to teach anatomy at Cambridge for a term. But the sense of how hard it must have been for my predecessors to make the original anatomical discoveries stayed with me. I was doing what they, the early anatomists, had done: cutting up dead bodies. But those plastic models in the doctor's office? It doesn't look like that at all. It's a confusing jumble of tubes and sinews. The question for me has stopped being 'how is the body put together?' and is now 'why is it put together that way?'

The early anatomists – the important Italians, for example: Eustachius, Vesalius, Malpighi – had to dissect, observe and catalogue with no knowledge of what the liver did or what a nerve was for. It is hard to imagine a contemporary equivalent. Perhaps mapping the outer reaches of the Solar System? Though the astrophysicists do not face either the smell of decomposition or illegality of obtaining scientific material that the anatomists did. They were in a very real sense uncovering secrets. Secrets that the church did not want them to know, that many older doctors did not want them to know and that the bodies themselves did not want them to know. No one in Renaissance Italy left their body to medical science.

When I want to feel better about my inability to tell one part of the body from another, I think of Leonardo da Vinci. Possibly the greatest draughtsman that ever lived and one of the greatest minds. A phenomenal observer of the human body in every way. And yet, he never entirely figured out how the heart and circulation worked. This is probably the only human organ whose function you can understand from simple macroscopic inspection. It still works after death to some

SECVNDA SEPTIMI LIBRI FIGVRA·

SECVNDAE FIGVRAE, EIVSDEMQVE CHARA-
cterum Index.

PRÆSENS figura sectionis serie primam subsequens, tertium duræ membranæ sinum (quem prima figura C aliquot insignium gerit) longa sectione secundum capitis longitudinem ducta adapertum commonstrat. Insuper ad huius tertij sinus latera, per capitis quoq; longitudinem duas deduxi sectiones, utrinque nimirū ad sinum singulas, quæ duram membranam duntaxat penetrarunt, & duræ membranæ latera ab ea membranæ separarūt parte, quæ dextram cerebri partem à sinistra dirimit, atque in subsequēti figura tribus D insignietur. Præter tres iam comemoratas sectiones utrinque aliā quoque molitus sum, quæ ab aure ad uerticē pertingēs,

folam

Andreas Vesalius (1514–64). *De humani corporis fabrica libri septum.* The last section of the *Fabrica* is devoted to the brain. Here, the dura mater has been peeled away, exposing the brain with its thin membrane and vessels. Vesalius drew such exquisite charts for his students that he became famous enough that the judges of Padua ensured a steady supply of cadavers from the gallows.

Chris van Tulleken, 4 October 2014. A picture of me holding a 500 ml pyrex flask of water, with blue food dye. Interestingly we seldom use blue liquids in this volume in my lab.

The human foot's navicular bone – it acts as a keystone at the top of the arch of the foot and is boat shaped, thus the name.

extent: if you go to the butcher's shop and buy a beef heart and fill it with water from the tap, it will still pump blood in the right direction if you squeeze it. But even the great da Vinci could not quite work out the order of valves and chambers such is the stringy, fleshy complexity of it.

Learning anatomy with the cadavers was a geography lesson: the rivers and mountains of the body all labelled and connected. It was a vast quantity of information. In the foot there are 26 bones. Taken out of the foot and held in the hand, they look like pebbles on a beach: they do not have a particularly discernible function. And yet assembled, you can see that they sit together the way that a stone bridge holds together: in your foot you have a keystone, the navicular bone. A miraculous example of biological engineering. I felt like I had acquired a lot of secret knowledge that first year, and all in Greek and Latin, so it felt doubly secret. I had begun to speak a language that my non-medical friends could not. And if you can draw the chambers of the heart and the valves and label the flow correctly, as most GCSE biology students can, then you're doing better than the smartest man in seventeenth-century Florence. In studying anatomy you feel like you are approaching a complete catalogue of the secrets of the human body: after all, if you know every road and house and place of interest in London, then you know London right? Of course not. There are other kinds of knowledge that are far more secret.

Doctors are in the business of keeping secrets. We keep them about our patients. All the details of a medical consultation, no matter how mundane, are confidential except in very unusual circumstances. Patients rarely confess to murder or plan to deliberately spread their deadly diseases so, although these potential dilemmas are popular among medical students determined to imagine that their careers will be difficult in exciting ways, the secret-keeping pretty much boils down to not talking about what you heard. Why is this so important? Because knowledge makes us vulnerable in many ways. Secrets must be kept not because they are illicit or shameful but because they can be exploited. Your business competitors, employers, insurers, bank and maybe even relatives are all in a position to exploit knowledge about your body: your fertility, your risk of future illness, your health fears and the limits of your abilities. Medical confidentiality isn't just about privacy, shame or discretion. It's about vulnerability to exploitation. That is why this book is not an anatomy book: the secrets we're interested in lie deeper. Your body is in the business of keeping secrets

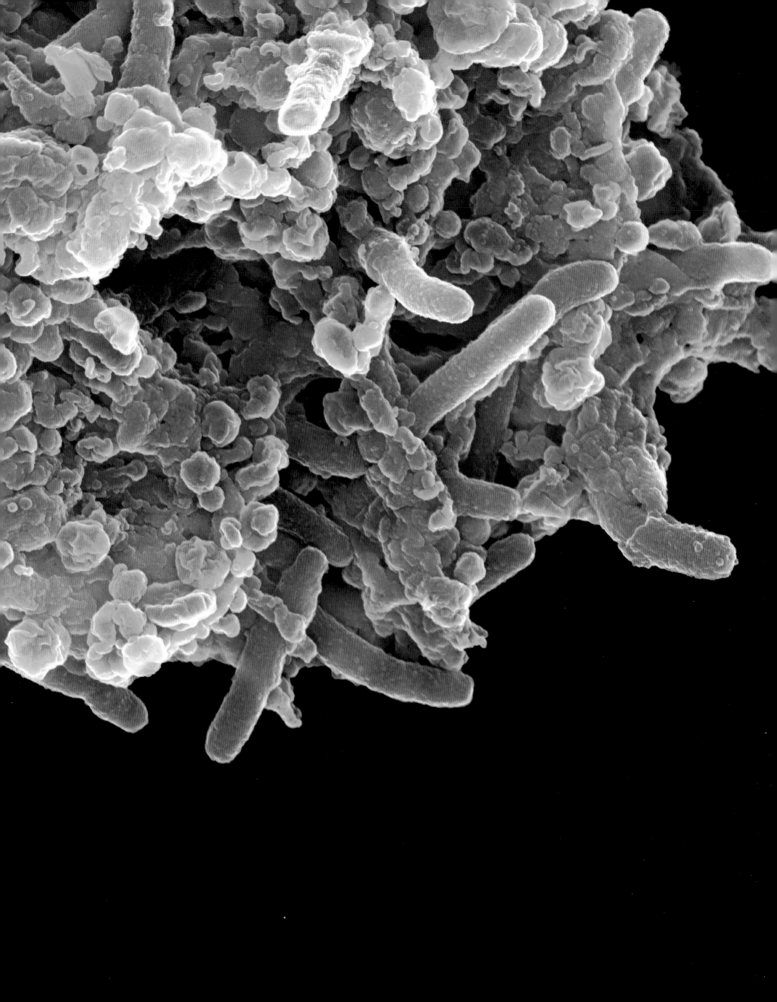

from everything that wants to exploit it: bacteria, viruses, fungi, parasites, larger predators and, crucially, other people. All these things are constantly probing our bodies, looking for weaknesses and opportunities. We survive by not giving anything away that we don't absolutely have to. This book is about those secrets, how we keep them and the people who uncover them.

The scientists who study the human body are not stargazers or map-makers, charting the features and dimensions of some territory in ever greater detail. They have to have the mindset of detectives or spies or tabloid journalists: they are digging for things that are deliberately concealed, information of life and death importance to those who keep it and those who seek it. This is what makes the stories in this book so thrilling: they are about stuff we are not meant to know. The important scientific discoveries about our bodies have both the deliciousness of gossip about who slept with who, and the heft of state secrets about where the nuclear submarine fleet is stationed. Let me explain.

There are two things you have to understand if you want to expose a secret. First, that a secret is a thing that is known, but only to a few. Secrets are not simply mysterious things that we can't explain; they aren't just obscure facts, or stuff that's too complicated to understand. They are hidden, deliberately, and they are 'knowable'.

The second important thing to understand about a secret is that it is a hole in the truth. A missing piece of a jigsaw. We notice secrets in their absence: a non-explanation; a story that doesn't make sense.

The last secret I was involved in keeping was pregnancy: my son was on the way. Like most people we kept it secret from just about everyone until the 12-week scan. But to keep this secret you can't just not mention that you're pregnant. You'll need to conceal or explain changes in your habits. So if people are watching you closely and consider your age, relationship status, recent weight gain and refusal of blue cheese and booze at the company picnic, they won't struggle to figure out the missing piece of the puzzle. To keep a secret you either have to lie or conceal a much wider array of facts. Whether you're pregnant and trying to keep it to yourself, or you're a government intelligence agency hiding their knowledge of the location of that fleet of nuclear submarines, the tactics must be the same: dissemble and confuse. And the tactics of anyone wanting to discover either secret take this into account: gather as many of the facts as possible until you can see the exact shape of the hole in the truth. Only the secret can fit in that hole. This is how much of biological science works. Using the facts you know to tell a story and then seeing if new facts are consistent with that story and revising it to make them fit.

The human body keeps secrets for the same reason that companies and states

Coloured scanning electron micrograph of *Mycobacterium tuberculosis* bacteria (purple) and a macrophage white blood cell. *M. tuberculosis* has the remarkable capacity to survive inside the very cells that should destroy it and is still responsible for around 2 million deaths around the globe each year.

and you, as an individual, keep secrets: we live in an environment of relentless competition and exploitation. All of life is exploitation. In most cases this exploitation is about who eats who; occasionally about who eats what. This might sound dramatic if you live in the UK. We don't seem to be in competition with much of what we eat or at much risk of getting eaten. But we all have an extensive set of defences to prevent us being eaten or exploited. Some of these things are complex behaviours, some of them work at a cellular level via antibodies and the killer cells of the immune system. But without all of them constantly functioning, we would be consumed within hours.

The relentlessness of the attacks on your body becomes clear when a person stops fighting them, even briefly. I used to work in the Bone Marrow Transplant Unit at King's College Hospital. In order to give someone a new bone marrow (this might be done because they have a bone marrow cancer like leukaemia), you have to kill their original one with radiation and chemotherapy, and this completely wipes out most of their immune system: they lose their ability to make antibodies or white blood cells. Sustaining life without a functioning immune system can be done for a short period of time. Our patients lived in positive pressure rooms, with minimal human contact and sterilised food. Even so they were often overwhelmed by infections and needed almost constant antibiotics and antifungals. They were, for an inexperienced junior doctor, some of the most terrifying patients in the hospital because they had no defence mechanisms of their own. They had to rely entirely on their medical team to keep them well.

We can also see the speed of attack on the defenceless body when you die. You switch off your resistance to being eaten and within hours you start to rot as your cellular immune system packs up and the bacteria take hold. And since your behaviours that should keep you safe – moving out of danger, for instance – also stop working at the point of death, your cat may start to eat you before the bacteria get a chance. Either way, the organisms that consume you have finally been given the break they've been searching for your entire life.

Every breath you take is filled with viruses, fungi, bacteria and, in many places in the world, nematode eggs (gut worms and the like). Every surface you touch is coated with other organisms. Every drop of seawater contains over 50,000 viruses. Your body is in constant battle: you endure wave after wave of assault every second of the day. It is in this fight that secrets become a matter of life or death.

So the body is a machine that has evolved to resist enquiry; to be inscrutable and unpredictable to those that would seek to exploit it. Throughout our entire evolutionary history, humans have been bombarded with organisms that would like to know much more about us: the limits of our genes, how fast we can run, the state of our antibodies are all valuable bits of information. Potential mates also want to know how ill you are or how long you'll live. If you know how far or how fast an organism can jump, or which molecules on your surface its immune system uses to recognise and destroy you, then defeating it becomes straightforward. Much of this information must be concealed for most of the time. This presents a challenge to anyone wanting to understand the human body, but it also presents an opportunity.

Many of our tools for probing the human body at the molecular and genetic level are stolen from other organisms that use them to probe us, or fight their own wars. Molecular biology laboratories might look like they are dominated by high-tech machines, row upon row of gleaming works of precise modern human engineering, but in fact the machines that sequence DNA or screen for cell surface markers or identify different biological molecules in different cells are entirely dependent on ancient biological materials: antibodies, enzymes and genetic fragments. The only way we are able to do genetic engineering is because bacteria and viruses have spent millions of years figuring out how to manipulate and exploit our cellular machinery to help them reproduce: they are the original genetic engineers. We could never design a DNA-polymerase enzyme – probably the most important single component of the genetic revolution – on our own. We have co-opted bacteria and viruses to be our double agents to make us stronger.

There are other barriers to discovering the secrets of the human body. If the route to uncovering a secret is through understanding what is known, and filling in the gaps, this is because the facts can be assembled into a coherent narrative. Those narratives are extremely hard to assemble in the case of humans because they played out in ancient history. We have genes that are millions of years old

Bacteria that live at high temperatures around thermal vents are a source of enzymes that are thermostable and have a wide range of applications in molecular biology. The best-known of these is the Taq DNA polymerase (structural model shown here), which is the foundation of the Polymerase Chain Reaction used for amplifying and sequencing DNA.

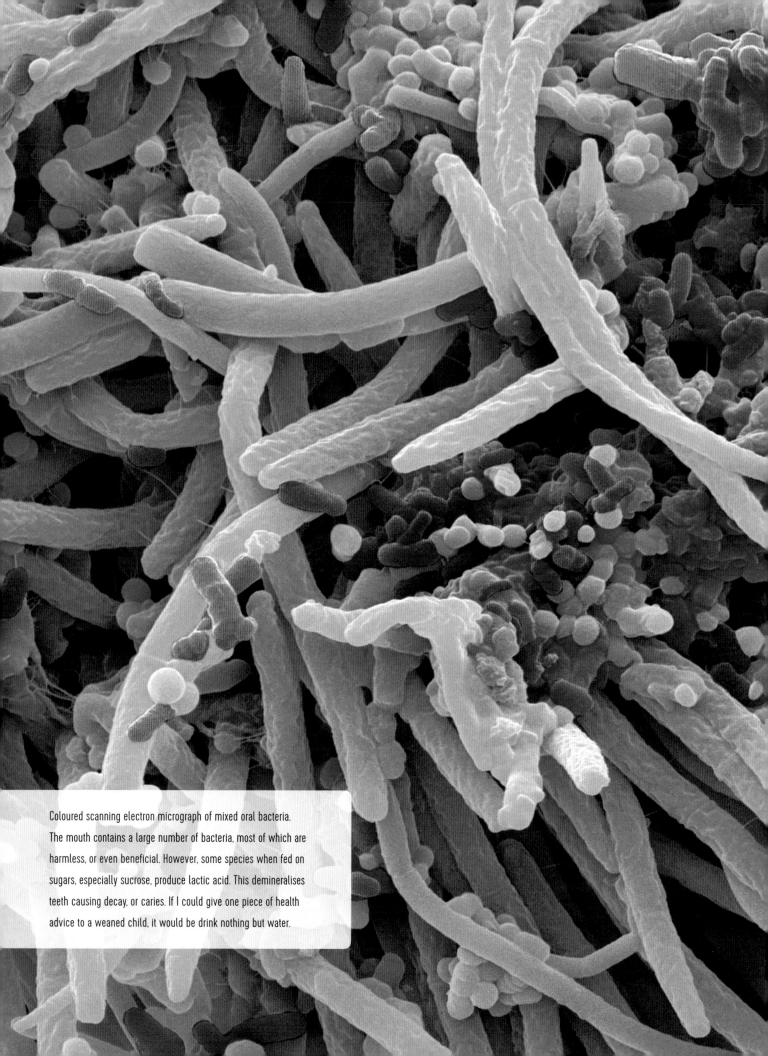

Coloured scanning electron micrograph of mixed oral bacteria. The mouth contains a large number of bacteria, most of which are harmless, or even beneficial. However, some species when fed on sugars, especially sucrose, produce lactic acid. This demineralises teeth causing decay, or caries. If I could give one piece of health advice to a weaned child, it would be drink nothing but water.

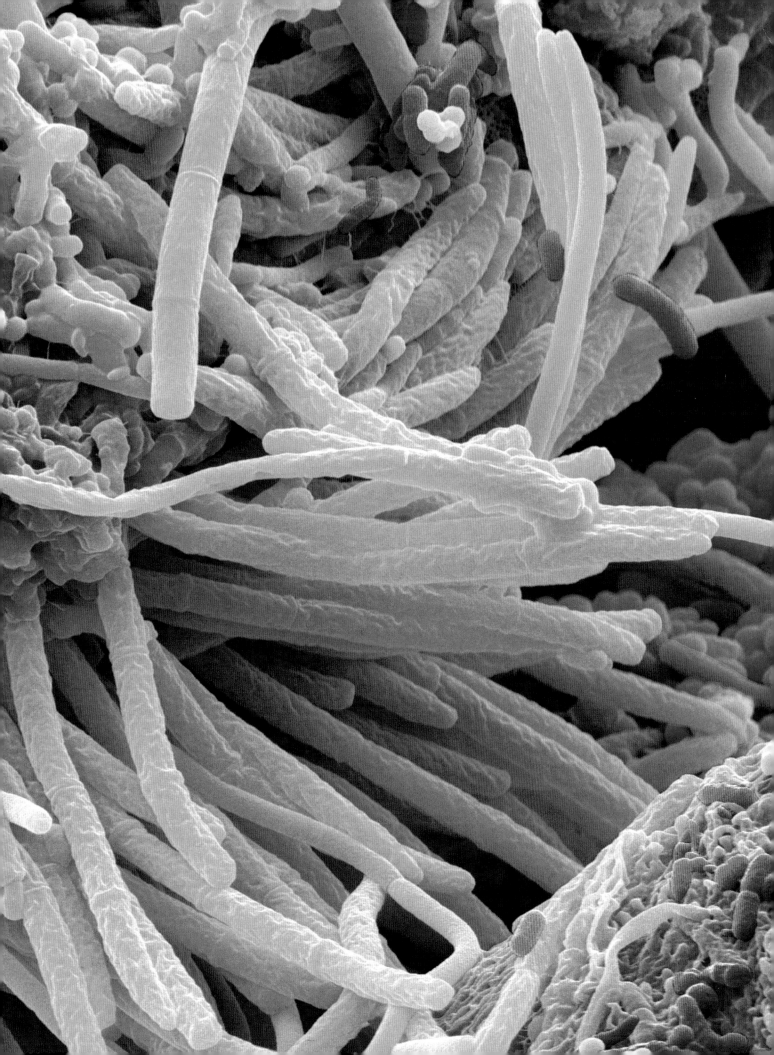

and our human bodies evolved under conditions that no longer exist for many of us. It would be a mystery why we have such an active anti-parasite system if we only considered life in contemporary Britain. But we evolved to co-exist with a vast parasitic burden: gut worms, liver flukes and malaria among many other invaders. An adaptation to resist a disease can only be understood if you know about the disease. Indeed, any aspect of the human body can only be understood in the context of the ecological niche we occupied millennia ago: our food supply, competitors, predators and environmental hazards. The world that we evolved in is largely gone and we face new threats now: old age, car crashes, high-calorie diets, sedentary lifestyles and others. Our bodies are designed to fight the previous battles in the same way that armies, their choices of camouflage, tactics and weaponry, frequently reflect the last war not the coming one. Our genetic code contains millions of years of alteration through mutation and selection. Each alteration adds caveats and subclauses to our genetic code, like amendments to the body's constitution: impossible to understand without an accurate knowledge of the circumstances in which these changes became desirable, much as it would be impossible to understand the laws of England without understanding the situations in which they were created. The right to move your sheep across Westminster Bridge for instance is unfathomable to the denizens of contemporary London. And yet the law still exists, a remnant of a previous time like some defunct part of our genetic code.

So if we want to understand the human body, we must understand it in the context of evolutionary history. This may seem straightforward: being able to outrun the sabre-toothed tiger (or at least run faster than the person next to you) will allow you to pass on your genes. Adaptations that allowed our ancestors to mate with more people and outcompete our human non-ancestors for food seem like valuable explanations for how we came to be the way we are. In fact the vast bulk of evolution is driven by a more universal phenomenon. The need to fight every organism in every square metre we occupy for what we call ecological capital (but can think of approximately as food and nutrients). Every place on earth only has a fixed amount of ecological capital. Some areas have more than others – the equatorial rainforests with their fertile soil and year-round sunshine have more than the Arctic or Antarctic – but for any given place it is fixed. This means that every kind of life from single-celled organisms to vertebrates is constantly evolving to try to get a little more. And so in order to survive we had to evolve too. Not just to beat the human living next door, but just to keep pace with the other organisms. This is is the Red Queen Hypothesis. It forms the foundation of much of the work done in Professor Greg Towers' Lab at UCL where Chris completed his PhD. It was

first described by Leigh van Valen, a towering genius who had to count a hell of a lot of fossils to demonstrate he was right. The Red Queen refers to the scene in *Alice in Wonderland* in which the chess board changes so rapidly that Alice must keep running just to stand still. It is an almost literal arms race: and no one ever really gets ahead (though quite a few species drop out and become extinct, while other species are created from the changes to take their place).

This vast accumulation of 'improvements' simply to keep up is not limited to our fight with other organisms. The extent of the complexity of the human body can be seen in a phenomenon described in a *Nature* paper in 2014 with the seductive title 'An evolutionary arms race between KRAB zinc-finger genes *ZNF91/93* and SVA/L1 retrotransposons'. What on earth does this mean? Well, we have mobile genes in our DNA called retrotransposons. They are old viruses that have inserted themselves into our DNA so effectively that we have inherited them for millions of years. Our bodies cannot be in the business of just replicating viral DNA for free. Otherwise we would become walking virus factories. That's the trouble with

John Tenniel's illustration of The Red Queen's race in Lewis Carroll's *Through the Looking-Glass*: "'A slow sort of country!' said the Queen. 'Now, here, you see, it takes all the running you can do, to keep in the same place. If you want to get somewhere else, you must run at least twice as fast as that!'" This quotation gave the title to Leigh Van Valen's 'Red Queen' Hypothesis that describes the continuous competition between organisms, where no one species ever gets ahead despite ongoing evolution.

■

# 'WE ALL CARRY IN OUR BODIES REPTILIAN GENES AND FISH GENES. FOR MOST OF THE ENZYMES WE MAKE, WE HAVE THE SAME SET OF GENES AS FISH.'

## SUSUMU OHNO

■

viruses: you give them an inch and they take a mile: these bits of mobile DNA can ruin our useful DNA. So we have developed genes called zinc-finger genes that act to bind to the retrotransposon DNA and stop it replicating. So far, so good. But remember we're in an arms race. They don't quit! These retrotransposons are forever adapting and breaking free of the zinc fingers. Over many generations they find a way to start replicating again. And so our zinc fingers expand to suppress them again. Just the way that when the cheetahs get faster to catch the antelope the next generation of antelope get faster too. We are in an arms race inside our own bodies against parts of ourselves. The lovely part of this – which seems on the face of it to be extremely annoying, like a rumbling civil war or secession movement – is that the technologies we develop to suppress the retrotransposons turn out to be useful ways of regulating other parts of our genome. Much in the way that military technologies can often have beneficial civilian uses: advances in aviation and radar and so on. Our internal battles make us stronger.

This is one of many examples of the vast complexity of our genome, which contains so much of our ancient past, and it creates a kind of fog of war. Complexity is an excellent way of keeping secrets.

There is a detailed explanation of DNA and the nature of our genome on pp. 137–9. But it is far from a simple blueprint. When it was decoded (one of the few vast public projects that arrived on time and under budget) it seemed we were on the brink of some new moment in science. We had finally been given the keys to the kingdom. But instead we were presented with further secrets.

The variation in genome sizes is bizarre: the lungfish, a particularly unglamorous and uncomplex vertebrate, has a genome 40 times larger than ours. It is reasonable to ask why is ours so small? (We don't know.) But it is also reasonable to ask why is it so much bigger than it apparently needs to be? We have vast quantities of genetic material that doesn't code for proteins. It doesn't seem to do anything.

It is reasonable to suggest that these sequences may have a function ... but that it is a secret. But in 1972 it was written off as 'junk DNA' – a term coined by Susumu Ohno, one of twentieth-century America's greatest scientists.

Susumu Ohno was a Japanese-American geneticist, and was said to have 'thought at least half of the thoughts' that form the basis of modern genetic research. He was showered with prizes and honorary degrees for his work on genetics and on his death in 2000 the Emperor of Japan sent his family condolences, a rare honour. His brilliance is not in question but he seemed to have an impish sense of humour. It may be that he used the term 'junk DNA' facetiously: he certainly seemed to believe that there were secrets locked up in our genes that might only be revealed in unusual ways.

In 1986 he gave an interview to the *Chicago Tribune* about his attempts to convert the human genetic code into musical form so that the vast repetitive sequences of code could be experienced in new ways. Whether or not this shed any truly useful light on the secrets of the genome, it does force you to encounter the rhythmic, repetitive nature of the code of life, and his arrangements have a surreal beauty that is hard to deny. In the same interview, he said the following: 'It is surprising that our ancient genes are not expressed more often. I think it's possible that babies sometimes are born with tails. But the surgeons just snip them off, and

The South American lungfish, *Lepidosiren paradoxus*. Lungfish were common in Devonian and Carboniferous times (400–300 million years ago), but only six species are known today. They are a group of amphibian-like, air-breathing fish, thought to be the ancestors of the terrestrial vertebrates. The South American lungfish has paired lungs, located on either side of the throat, and can survive for long periods if the river or lake in which it lives dries up.

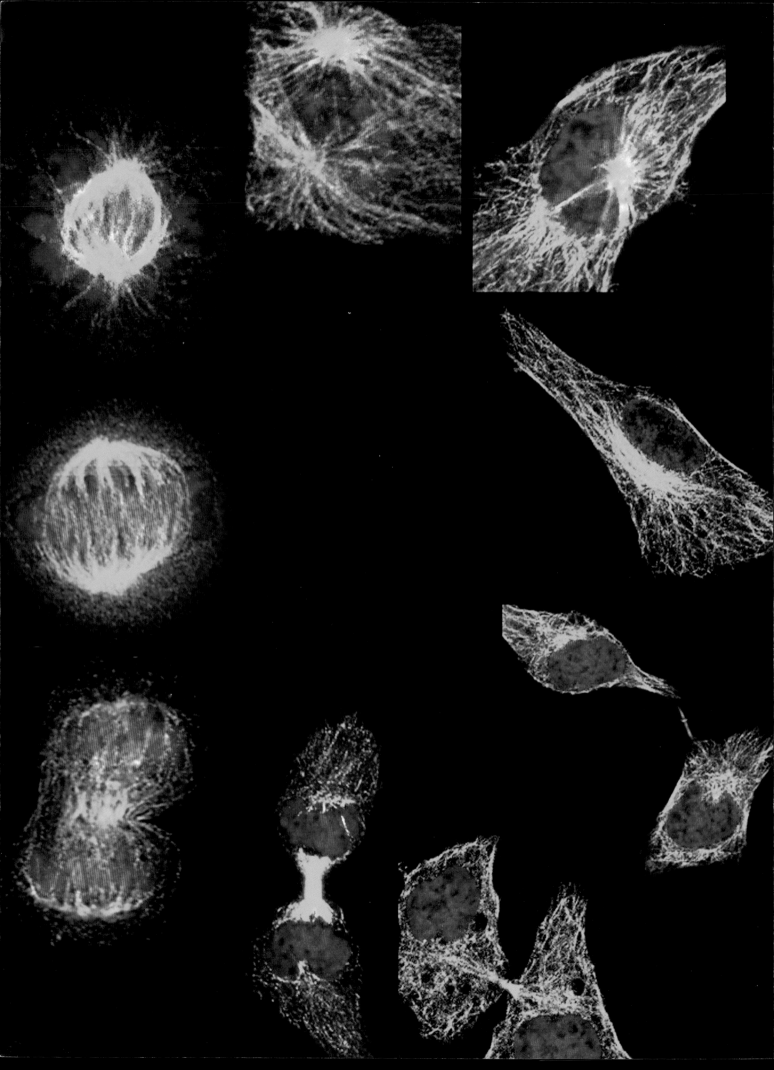

we never hear about them ... we all carry in our bodies reptilian genes and fish genes. For most of the enzymes we make, we have the same set of genes as fish.'

There is no medical conspiracy to conceal the tails we are born with – or if there is, I'm out of the loop – but it's a great line from a revered scientist and captures the bizarre accumulated nature of what is sometimes described as a conventional blueprint: the instructions to build a human. Our genome is more like the instructions to build a human, and also the instructions to build a ton of other ancient equipment and viruses, and also the instructions on how to prevent anyone using the wrong sets of instructions.

So it seems like it's not junk – the bundles of DNA allowing for complex evolution keep the secrets of our past. Even once we have the entire code in a computer database, as we do now, we barely understand its functions. This is also why this book is not an anatomy book: you can have a lot of facts, in this case all the facts, and still they can conceal secrets.

This book is arranged around three themes: learning, survival and growth. There is a chapter devoted to each. Every aspect of the human body is shaped by the need to achieve these things. Human bodies, like all organisms, have one central job: to ensure the survival of their genes. Learning, growth and survival are all vital to achieve this goal. Counterintuitively, these three tasks may be even more important than reproduction. Reproduction is a useful but not essential task for gene survival: you don't need to reproduce to ensure your genes survive. It helps because it puts more copies of them out there but if you look after your relatives, who share your genes, and help them survive and reproduce you are still giving your genes an advantage over your non-relatives. If you have an identical twin then your sibling will share all your genes, and their children will be – according to any genetic test – your children (my son Julian calls Chris 'Uncle Dad' and occasionally just 'Dad'). But you can't be any use to your family members unless you have managed to grow, learn and survive. These, the secrets of the human body, are the secret ways in which we all achieve these things in spite of the thousands of things that try to stop us every second of every day.

Growth, learning and survival are interconnected: they are each part of a complex network of processes and forces that make us who we are. And they cannot happen independently. As we learn our bodies grow and our survival is predicated on our ability to learn from threats and to learn how best to exploit our ecological niche.

Learning is not simply the acquisition of facts and memories. The 'Learn' chapter is about the incorporation – the literal embodiment – of the physical and social world into ourselves. Ulysses, in Tennyson's poem, reflects on his life and his worn-out body and his future:

The moment of growth. A cell divides. False-colour transmission electron micrograph of cytokinesis – this is the moment one cell becomes two, the red genetic material having been doubled and divided. It is the final stage of mitosis or cell division.

I am a part of all that I have met;

Yet all experience is an arch wherethrough

Gleams that untravelled world whose margin fades

Forever and forever when I move.

There is an ambiguity here: our bodies imprint themselves everywhere we go and our lives are incorporated into us. We shape the world as it literally shapes us. Our brains lay down our experiences not through some ineffable neural magic but with visible layers of fatty myelin – white matter – coating our nerves. In the 'Learn' chapter we will see how scientists are now able to watch our memories – or the proteins that write them in our brains – travel the length of neurons to be stored for retrieval. And we will see how we can track the timing and locations of our short- and long-term memories: the different processes by which we transiently remember a name at a party and store the feeling of our first day at school for the rest of our lives. We will meet unique people, each of them a Ulysses of their own world, exploring and pushing the boundaries of human experience: Danny MacAskill, the uniquely skilled cyclist; Deb Roy, the scientist with (surely) the largest home video collection in the world; Akash Vukoti, the youngest ever participant in the US spelling bee championships with a vocabulary that would shame most adults; Henry Molaison, the man who lost the ability to make new memories. These people in these stories will show us how our brains and therefore ourselves are shaped by what we learn. Ulysses describes experience as an arch, a robust yet delicate structure built of the world through which we travel. Our bodies are literally built of our experiences. They learn and adapt to the untravelled world that faces all of us: our bones are shaped by the forces we experience: in the chapter we will meet astronauts and tennis players who are broken down and rebuilt by microgravity or the repeated impact of ball and racket and we see how the world writes itself upon us. Like Ulysses we never stop moving or changing: our bodies are designed to learn from experience and adapt to threats we can have no knowledge of until we meet them. Our ability to learn is what allows us to move into the untravelled future: I may never 'drink delight of battle with my peers, far on the ringing plains of windy Troy' but we are all constantly learning to overcome the particular challenges life throws at us. And don't worry, you can teach an old dog new tricks: we will see Chris valiantly attempting to prove this as he goes up against an 8-year-old in a juggling competition.

Parts of growth are about learning: we grow stronger, or we grow tougher in certain ways as we learn from the environment. But the growth chapter centres around the most striking aspect of growth: a typical human will increase its size by over 20 times from birth to adulthood and this vast increase must occur without interrupting learning or jeopardising survival. Growth is extraordinarily energy expensive and most of this energy in the early and most rapid phases of growth comes entirely from breast milk. Breast milk is the only stuff on the planet that has evolved specifically to feed humans, it has determined our ability to grow through our entire evolutionary history and yet we have only just managed to understand the role of its main ingredient.

We will meet one of the tallest families in the world and through them examine what drives us upwards, what advantages and disadvantages it might confer, and see the extraordinary mechanics of growth, the architectural equivalent of building a miniature building and then expanding every part of it over the course of 20 years while constantly improving its function. This remarkable growth occurs in two spurts during childhood and adolescence, and we will see how the demands of growth must be balanced with the immediate demands of learning and survival. But even once we have reached adult size we do not stop growing. We will meet Lew Hollander who, at 87 years old, is still competing in Iron Man triathlons and making demands on his body that require new growth. We will see how he forces us to consider the role of wear and tear in stimulating growth and the way in which the body never stops growing.

The fact that we never stop growing – that our cells have the ability to divide trillions of times – provides an enticing opportunity: the possibility of human growth without a complete human body. We will meet Professor Harald Ott who is growing human hearts in his laboratory. If he succeeds it will be an almost unprecedented milestone in the history of medicine and we will see how much his work relies on one of the most important but uncelebrated parts of the body: the extracellular matrix. This is a lattice of proteins and sugars that tells dividing cells where to sit and what to do. It binds us together so that we are not simply slime. We will see how the architecture of the heart allows it to pump blood so efficiently and unfailingly as it grows and also see how the forces that the heart itself generates are essential in its growth and function.

Learning and growth are fundamental to our survival and to the survival of our genes, allowing us to repair damage, reduce threats and learn from previous encounters to interact with our environment in a more sophisticated way, and the 'Survival' chapter presents the challenges of doing this. Why must we learn and why must we grow? Because we are so delicate. Because we are unable to withstand even the smallest changes to our internal environment. But we need to have innate mechanisms to protect us from the vast variability in the world because there is so

A range of glassware with food dye and water. Even twenty-first-century science requires huge quantities of clean glass. In my lab I use flasks like the large two-litre ones on the right of the image to grow *E. coli* bacteria for genetic engineering. I always wear a satin bowtie in the lab. Xand is just copying me (he doesn't work in a lab).

little margin for error and we need ways of incorporating this variability into our behaviour and reactions to threats.

We begin the chapter with Chris witnessing the moment of conception in a Harley Street IVF clinic. The miracle of this moment is almost overshadowed by the miracle of human homeostasis: our ability to keep our internal environment constant in almost every way. That cell will not change temperature, pressure, acidity, oxygen concentration or anything else until its owner dies. And at least one of its owner's cells will endure in that same environment indefinitely as long as they have a direct line of descendants.

Of course scientists are rarely happy to simply agree that the internal environment of the human body is pretty consistent. They want to see just what it is possible to endure in the way of external changes. And so Chris and I went to Professor Mike Tipton's extreme physiology lab in Southampton where we were taken on a journey from the high Arctic to the desert to see just how much temperature variation our bodies could handle.

It is easy to imagine as a modern human that we live our lives in rational ways, our decisions based on education and experience. We have brains developed to cope with very different times and parts of our brains are very old indeed. We will see how the more recently evolved parts of our brain govern the older more instinctive parts of our brain. We live in a delicate balance with emotions like fear and disgust that serve to protect us but simultaneously can potentially disable us. Disgust is particularly complicated. It is our least-considered emotion, but most of the time our disgust sensors are turned up full volume. Just occasionally we have to turn them off entirely, to reproduce or eat. Chris and I encounter the most disgusting meal we have ever eaten and realise that food and sex are linked in ways you might never imagine (... and might prefer to continue not imagining for the sake of both your love life and your dinner table). We tour through fear: experiments we are no longer allowed to do show how our bodies fine-tune our sense of what to be afraid of. And we learn about the one thing we are all born frightened of and what happens if you completely lack the capacity for fear.

This book isn't just a catalogue of the intriguing or miraculous. We want to show you the secrets to the way that the human body is interrogated, the way in which it gives up information agonisingly slowly and reluctantly. Much of writing this book, and making the accompanying television programme, felt like bring-your-child-to-work-day. Chris and I got to be naive and to ask simple questions of amazing people. Questions like 'why does it do that?' and 'why is it made that way?' are the sorts of thing children ask but when you pose them to the world's best scientists you get extraordinary answers. We were able to arrange absurd scenarios like having the most disgusting dinner party with a scientist we had only just met and persuaded colleagues to torture us to make a point about homeostasis. We got to watch a baby get made. Why isn't this just a textbook? Because this mad cascade of events and facts needs meaning. We wanted to give you a way of thinking about yourself, and to let you in on some of the secrets that your body has been keeping from you.

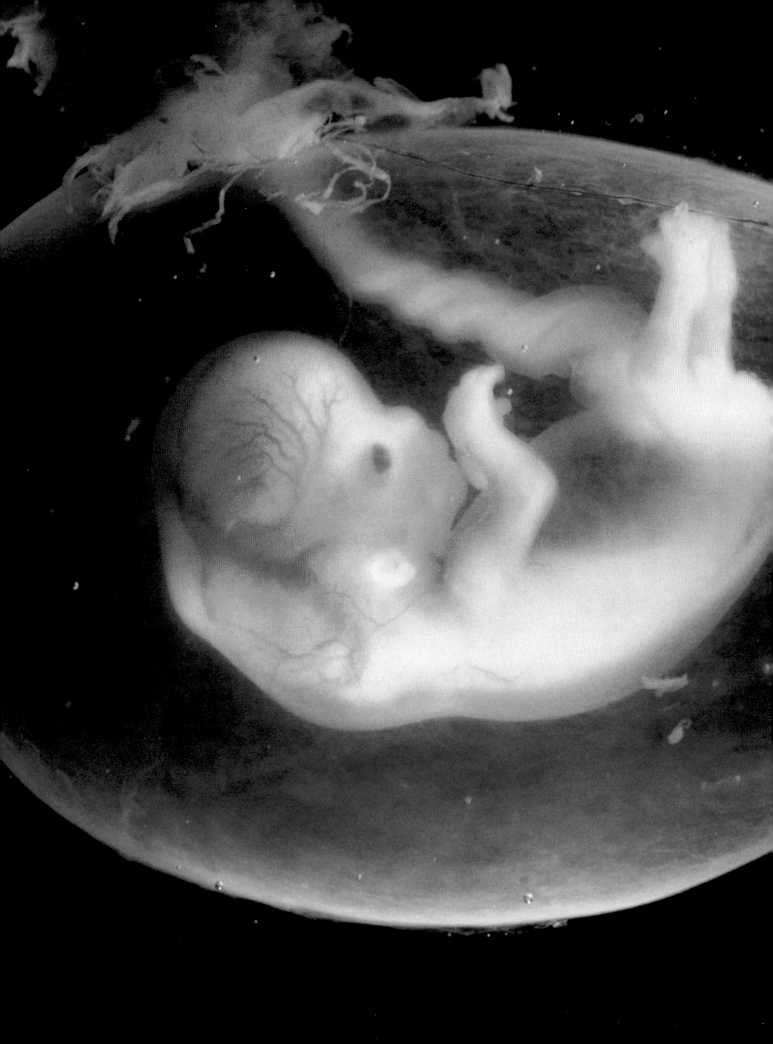

ONE
---

# GROW

# BABY TO BABY-MAKER

The simple process of growing is an extraordinary thing. During a lifetime our growth rate is truly staggering: from starting out as a single fertilised egg at the moment of conception, we multiply into a mass of trillions of cells made up of over 200 different cell types organised into around 80 organs (the exact number depends on how you define an organ ... not as straightforward as you might think!), and that's before we are even born.

At birth the average human weighs about 3.5 kg and is approximately half a metre in length from head to heel. On our journey from baby to adult we then go through an amazing transformation: we quadruple in height. We add, on average, 70–80 kg to our body weight, although far more than this is increasingly common. At our fastest rate of growth we can elongate by up to 1.5 cm in a single day. It's a process that we hardly notice but exploring how, why and when we grow reveals an extraordinary secret inside our bodies, which ultimately leads to a transformation that every one of us goes through. In this chapter we will explore the latest understanding of that process of growth and the magic ingredient that fuels it through the beginning of our lives. We will uncover the mystery of human childhood, a childhood longer than any other creature on earth, and explore the mysterious

From babies to baby-makers.
**Top:** Chris's wife, Dinah, swears that Xand is on the left, but no one knows for sure; **Below left:** Chris with Lyra. **Below right:** Xand with Julian (and yes, you spotted it, Xand is nude).

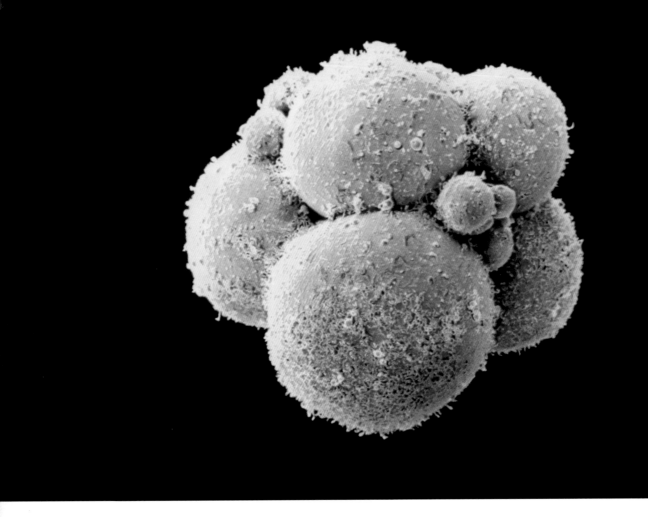

Coloured scanning electron micrograph of an eight-celled human embryo three days after fertilisation. Known as a morula, this is a cluster of eight large rounded cells called blastomeres. Smaller spherical structures (centre right and left) will degenerate. This embryo is at the early stage of transformation into a human composed of trillions of cells, or two humans over trillions of cells (in the case of twins). At this eight-cell stage the morula has not yet implanted in the uterus.

moment that triggers the body of a child to suddenly transform itself into an adult. We willl also discover that growing doesn't just end at adulthood, because throughout our lives our bodies are endlessly replenishing and regenerating, and even in old age, in some ways, we still continue to grow. We are now using this knowledge at the very cutting edge of medical science, to redefine our perception of human growth by learning how to replicate it, control it and ultimately build new human organs and tissues grown entirely in the laboratory.

I was born at 13.45 on 18 August 1978 with Chris taking an extra seven minutes to emerge into the welcoming arms of a midwife at the Queen Charlotte Hospital in London. At birth I weighed in at 6 lb 12 oz (3.06 kg) with Chris a slightly chubbier 6 lb 14 oz (3.12 kg). Thirty-nine years later and the vital statistics have not played out in my favour. Chris is not only an entire half an inch taller than me at 6 ft 1 in (185 cm) but he is also from the last available records approximately 5 kg lighter than me as well. Small differences to you perhaps but when you're an identical twin it's these differences that really matter! But the changes in our height and weight are just two of the miraculous transformations that we have gone through over the last 39 years. Each one of us sees our body transform throughout our childhood and beyond under the influence of a multitude of different factors, a complex web of genes and environment that combine together to turn a baby into a baby-maker. As we'll see in the 'Learn' chapter, the transformation of our brains from newborn to highly skilled adult is a miraculous journey of its own, but our bodies undergo an equally extraordinary transformation. The size, shape, strength, appearance and function of our bodies are completely altered through those first 18 or so years of our lives. It's such a gradual process that until we compare young and old photographs we often miss just how comprehensive and extreme a physical change this is.

To put the extraordinary nature of this process into a slightly different context, just imagine attempting to build a machine that has such an inherent ability for self-transformation in both its structure and function adapting constantly to the demands made on it. In engineering terms it would be a plane strengthened by turbulence, a car that got faster the more you drove it and used less fuel per mile, a computer that got quicker and more accurate with age. On top of that these machines would be able to fuel and reproduce themselves. As we will see in this chapter, attempts at bioengineering reveal just how superb and precise a designer nature is compared to even our most impressive innovations.

For most animals on the planet the journey to maturity is a quick and efficient process. Many mammals, including dogs and cats, reach adulthood within six months of birth, blue whales (the largest animals on the planet) are able to reproduce as quickly as five years after birth and our nearest cousin the chimpanzee completes its journey to maturation years ahead of us, being fully grown, sexually mature and reproducing on average by the age of 13. It seems an elongated childhood is a uniquely human process. No other animal on the planet takes longer to reach maturity and no other animal goes through such a convoluted stop–start process of growth and development than a human child. In many parts of the developed world the average age of a first-time mother is well into her thirties. Many of the reasons behind this elongated adolescence are of course cultural, heavily influenced by the way we structure our societies and the growing balances between the lives of men and women. But underneath the social influences there is an intriguing biological story with mysteries that we are still trying to understand. Why for instance do we have two discrete growth spurts, one just after birth and one on average more than 10 years later just before puberty? What is the reason for

**SELF-REPLICATING MACHINES**
A quick interesting side note whilst we are talking about this is that the concept of self-replicating machines has a long history in both fact and fiction with perhaps the most famous being devised by the Hungarian mathematician, scientist and general genius John von Neumann. Von Neumann machines that could explore and colonise whole galaxies have been the subject of much conjecture for decades and the fact we have never found one has been used by some as evidence that advanced civilisations are absent from this and perhaps every other nearby galaxy.

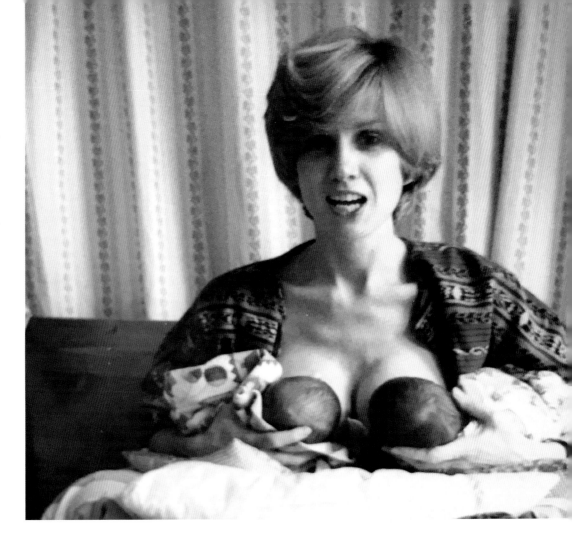

Breast-feeding twins. Our actual mother, Kit, breaking herself down to build us. She doesn't know that we've put this picture in.

this elongated lull in our growth? And what triggers the sudden onset of puberty? All of these are questions that until very recently we have struggled to answer but in the last few years many of the secrets of the journey from baby to reproduction are beginning to be revealed.

# MIRACLE MILK

There is only one substance on earth that is specifically produced as a nourishing food: milk.

A few fruits have evolved to be palatable, persuading animals to eat them and distribute seeds in faeces, but these are just sweet-treats. And yes, we can extract nutrition from animal flesh and a handful of plants and fungi, but our relationship with these foodstuffs is more complex, more competitive. They didn't evolve specifically to be food. Milk, and only milk, did.

Breast milk is what makes mammals, mammals. There are other characteristics that most mammals share but the platypus, and a few of its Australasian friends, mess things up, with their egg-laying and lack of placentas. But even the platypus makes milk.

Milk is extraordinary stuff. It is, most obviously, a complete source of nutrition. It contains fat, protein, carbohydrate, water, vitamins, minerals, amino acids and fatty acids, all in an available form, tailored perfectly to each stage of development. You can build a human child for several years entirely on breast milk. But it's not just food. It also contains an immune system in the form of antibodies, and a cocktail of hormones and other factors that regulate and stimulate infant development. And far from being a single substance, it is constantly changing, as the child grows, between breasts and throughout the day. Even within a single feed the milk shows complex fluctuations in what it contains in terms of lipids, carbohydrates and total calories.

Breast milk is made when a set of genes are turned on in the cells of the mother's breast during pregnancy. These genes encode proteins, including the enzymes that turn the mother's body into the end product. The genes for milk production

Lyra van Tulleken getting to grips with breast feeding at age 5 hours. At the time of publication she has yet to fully master it but her mother is more determined than she is.

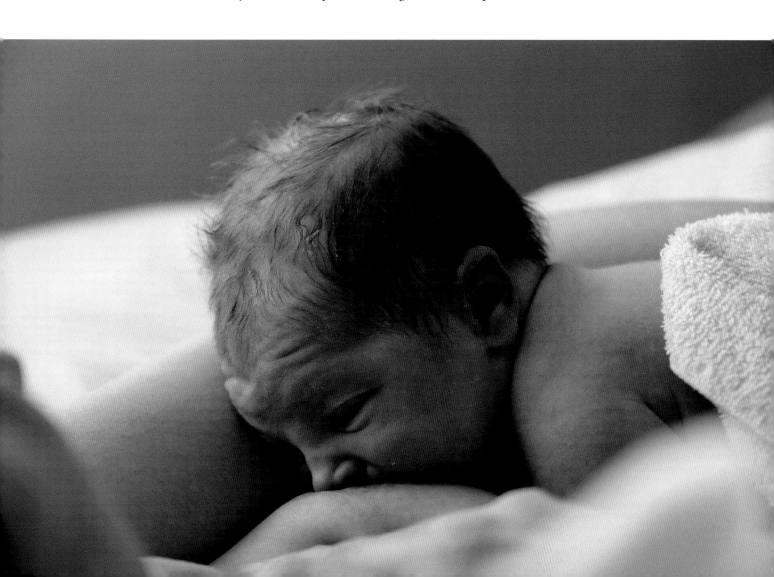

have been selected by evolution over around 150–200 million years since the first shrew-like creatures gave their young primitive breast milk from barely modified sweat glands. Yes, for all its erotic and maternal associations, the breast is a modified sweat gland. It's impossible to know what that first milk, produced somewhere around the late Jurassic era, would have consisted of, but considering that modern shrews can barely get enough calories to sustain themselves for more than a few hours, that early milk may have been more about the transfer of antibodies, to fight infection, than calories.

## THE COST OF A PINT OF MILK

For Bruce German, a chemist at the University of California, Davis, milk was the obvious starting point to understand nutrition. Nutritional science made a few huge leaps early in the twentieth century, before stalling around the 1970s. It had been established that to stay alive, we need three macronutrients (carbohydrate, protein and fat) and all but invisible amounts of a few vitamins,

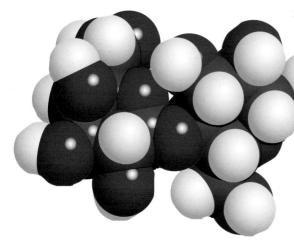

Model of a lactose molecule: carbon atoms are dark grey, oxygen red and hydrogen white.

minerals, amino acids and essential fatty acids. Biochemists figured out the chemical reactions that are required to turn what we put into our mouths into flesh and energy. They discovered the enzymes that enable these reactions to happen. They described how the molecular constituents of our cells are recycled and replaced in response to the world around us. But the ideal human diet continues to elude detailed description. There are some broad brushstrokes that we're confident about: eat lots of plants. Meat seems to be OK in small amounts. Fish is good. Refined sugar may be bad. But dig a little deeper and confusion reigns: saturated fats were bad, then good, then bad. Oily fish, vitamin supplements, low-carb vs. high-carb? These questions still generate inconsistent answers.

'Milk offered the opportunity to take an evolutionary perspective. What food *should* we eat?' says Bruce. It's worth saying that there is a lot of bad science based on an evolutionary perspective. It usually involves scientists discovering something about the way people behave towards their mates, friends or enemies and then inferring causality from hypothesised ancestral sabre-tooth tiger encounters. This is often not very useful because we don't understand much about how we used to interact with sabre-tooth tigers (probably very little). Bruce German and his team at UC Davis do not involve sabre-tooth tigers in their evolutionary perspective. They study the genes, enzymes and constituents of milk. Not all genes in the human body are treated equally by evolution. There are many ancient genes that remain stable, barely changing over long periods of evolutionary time. But the genes involved in breast milk production have been under intense evolutionary scrutiny since it first evolved. Because while milk is great for the offspring, it's pretty bad for the mother. This is because it is massively expensive for her to produce. And this was the reasoning that the team at UC Davis started with: human breast milk has evolved to be perfect nutrition for a human infant but it needs to be extremely efficient because it is so costly for the mother.

How costly? A mother will, on average, make about 750 ml (almost a pint and a half!) of breast milk per day for the first five months after birth. This will gradually increase with demand to almost a litre a day if exclusive breast feeding continues, assuming the mother is herself well-nourished and hydrated. As drinks go it is rich in energy containing around 65 calories per 100 ml. Unsurprisingly this is about the same as whole milk from a cow. By comparison a sugary cola drink will have about 40 calories per 100 ml. The cost to the mother is immense.

A palmitic acid molecule: the major saturated fatty acid in human milk, usually represents about 20–25 per cent of human milk fatty acids.

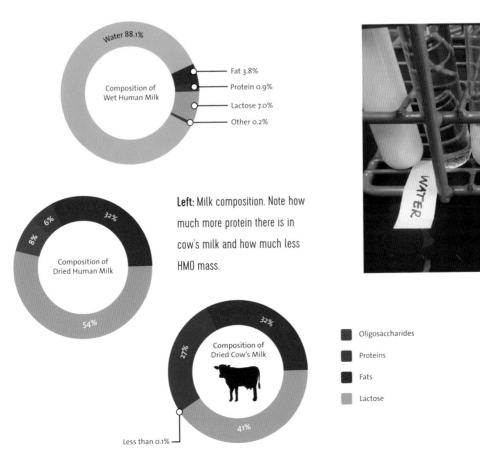

Composition of Wet Human Milk

Water 88.1%
Fat 3.8%
Protein 0.9%
Lactose 7.0%
Other 0.2%

Composition of Dried Human Milk

32%
6%
8%
54%

**Left:** Milk composition. Note how much more protein there is in cow's milk and how much less HMO mass.

Composition of Dried Cow's Milk

32%
27%
41%
Less than 0.1%

■ Oligosaccharides
■ Proteins
■ Fats
■ Lactose

**Above:** Components of human milk extracted in Bruce German's laboratory. The Human Milk Oligosaccharides are on the far right and are about the same in volume as the protein, indicating their importance.

A MOTHER WILL, ON AVERAGE, MAKE ABOUT 750 ML (ALMOST A PINT AND A HALF!) OF BREAST MILK PER DAY FOR THE FIRST FIVE MONTHS AFTER BIRTH.

**Opposite page:** *Bifidobacteria*, coloured scanning electron micrograph. *Bifidobacteria* are gram-positive anaerobic bacteria that live in the gastrointestinal tract, vagina and mouth. *B. Infantis* species have evolved to metabolise the complex oligosaccharides in human breast milk and confer a range of benefits on a growing child.

She will produce milk by breaking down her own body. Even if milk was made 100 per cent efficiently, it would still be a huge number of calories stripped from the mother, but to calculate the true cost you have to first work out the efficiency of milk production. And it's not a trivial calculation. Experiments have been done all around the world using isotope labelled water, special metabolic chambers and biochemical calculations about milk composition. Estimates vary but to make 1 calorie of milk takes about 1.2 calories of energy from the mother. All this adds up such that, on average, exclusively breast feeding a 6-month-old child will demand a large burger's worth of energy from the mother each day. And I mean a proper 650 calorie burger. So that's with bacon. And cheese. For the vast bulk of mammalian evolution obtaining this amount of energy came at a huge cost. It costs the mother her own body but there is also the evolutionary cost of the risks required to obtain nutrients. Foraging isn't just exhausting, it increases the risks of being eaten (but probably not by sabre-tooth tigers).

All this told Bruce and his collaborators two things about breast milk. Firstly, since you can grow a healthy human for many years exclusively on breast milk, it will have everything that a baby needs. Secondly, there is not likely to be anything in it that the baby doesn't need. From the moment the earliest mammals started producing milk, any mothers that wasted energy on putting unnecessary stuff into it would have quickly been plucked out of the gene pool. The solid components of milk that cost the mother most of the calories by order of amount are 1) fat; 2) sugar; 3) complex chains of sugar molecules called Human Milk Oligosaccharides or HMOs; and 4) protein. Each of these must be of absolutely vital importance to the infant. But here's the bizarre thing. The third largest solid component of milk, those Human Milk Oligosaccharides, are totally indigestible by a human infant. More than bizarre, it seems absurd. In the words of Bruce German, 'the mother is expending tremendous amounts of energy to produce these varied and complex molecules and yet they have no apparent nutritional value'.

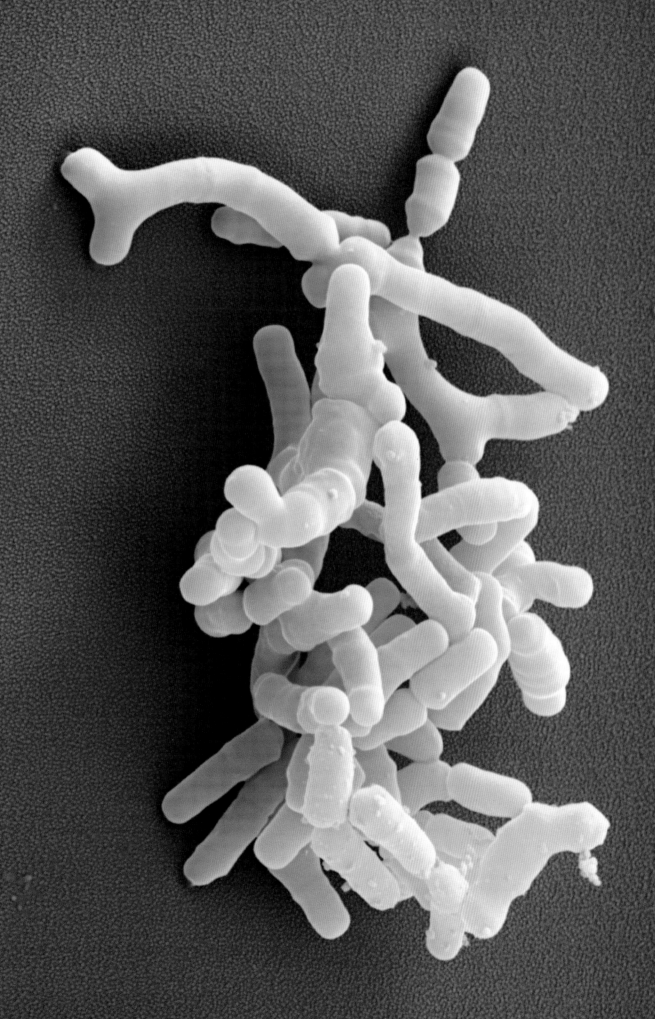

Human Milk Oligosaccharides are chains of sugar molecules. To put that in context it may be useful to understand a little about different sugar molecules. Monosaccharides are single molecules, usually rings of carbon with a few hydrogen and oxygen atoms added on. Glucose is a familiar example. As a single molecule, it can be absorbed into cells and used for making energy without any breaking down in the gut. Disaccharides are made of two molecules. The white refined sugar in your kitchen is a disaccharide called sucrose, made of a glucose molecule joined to a fructose molecule. The chemical bond that joins the two molecules needs to be broken down by enzymes in your gut before you can use the individual sugar molecules for generating energy. Polysaccharides are long chains of 200–2,000 sugar molecules. They're often indigestible, like cellulose. Oligosaccharides sit in the middle. The ones in breast milk are branched chains of between 3 and 22 sugar molecules with unhelpful names like *di-sialyl-lacto-N-tetraose* and *lacto-N-fucopentaose V*. There are around 200 unique and different types of oligosaccharide in human breast milk, each with different sugar molecules joined together in different chains. Crucially these all need different enzymes to digest them. And humans have none of them. We know that because, in the words of Bruce, if you feed a modern American child human breast milk, 'the HMOs come out the other end'.

So why is there the same amount of these totally indigestible oligosaccharides as there is protein in human milk? To feed bacteria. In fact to feed a single bacteria: *Bifidobacterium infantis.*

## BUG FOOD

The idea that the HMOs might be present to feed bacteria rather than humans is an old one. Over a century ago paediatricians, microbiologists and chemists were already trying to understand the health benefits and constituents of breast milk. In the last part of the nineteenth century in Europe and America one in three children died before the age of 5, but it was clear that the chances of survival were higher for breastfed infants. By 1900 Austrian doctors and scientists had detected differences in the bacteria found in the faeces of breastfed compared with bottle-fed infants, a remarkable achievement considering the technology of the time. As early as 1888 sugars other than lactose were identified as being in milk, and by 1926 it was reported that there were factors in human milk to promote the growth of a genus of bacteria called *Bifidobacterium,* but the extraordinary details of the relationship between human mothers and these bacteria took almost another century to determine and required huge advances in genetics and microbiology.

## THE HUMAN MICROBIOTA

Estimates for the total number of bacterial cells found in association with the human body have varied between 10 and 1.5 bacteria for each and every human cell. The total number of bacterial genes associated with the human microbiota could exceed the total number of human genes by a factor of 80 to 1. Conservative estimates suggest that an average 70 kg human being is composed of about 30 trillion human cells ... and 40 trillion bacterial cells.

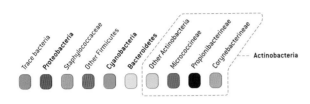

## MAP OF HUMAN SKIN BACTERIA

The human body provides a rich and varied environment for bacteria. Different parts of the body host very different communities of prokaryotes. Most of the microbes appear not to be too harmful, and many assist in maintaining processes necessary for a healthy body.

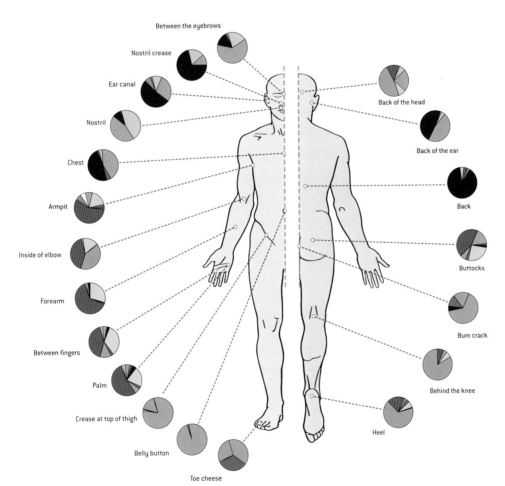

The human body provides a rich and varied environment for bacteria with different parts of the body hosting very different communities. In the right location the bacteria perform useful functions, but the concept of 'good and bad bacteria' is simplistic. The crucial thing is to have the right bacteria in the right place. Mouth bacteria on a heart valve are bad. Gut bacteria in the urinary tract are bad.

**Opposite page:** Lyra van Tulleken undertaking her first science project with her *Bifidobacterium* teddy (notice the characteristic Y shape). Beside her is a sachet of *B. infantis* probiotic and a box of faecal sample collection tubes to monitor whether her gut is colonised by the bacteria. Her data will be used for research by Professor Bruce German at UC Davis, in California.

There are a wide range of *Bifidobacteria* species but they all have a bifurcating shape under a microscope. Aside from that, distinguishing them all is not easy. Starting with the hypothesis that Human Milk Oligosaccharides would nourish them, Bruce German, together with David Mills, a microbiologist, started testing different bacteria, including multiple *Bifidobacteria* species, to see if they could be cultured in the laboratory using Human Milk Oligosaccharides as their only source of food. Surprisingly initial tests showed very unenthusiastic growth by any of the species tested. They seemed to lack the enzymes necessary to digest the wide variety of sugars in breast milk. But then the team tested *B. infantis*, a bacteria first isolated from the stool of a breastfed infant, and it flourished.

If digesting HMOs required a single enzyme, then this ability could be put down to coincidence. Perhaps *B. infantis* had evolved to digest similar molecules in other environments. But digesting HMOs requires a vast toolkit of genes. *B. infantis*, and only *B. infantis*, has them all. An analysis of the genome of the bacteria revealed over 700 unique genes compared to other *Bifidobacteria*. These include genes for grabbing the HMOs and taking them inside the cell, as well as a series of enzymes able to break down the full range of linkages between the sugar molecules. They are the only bacteria able to completely break down HMOs and there can be no doubt that they have evolved to do this. More importantly the evidence from other species shows a process of co-evolution. As the bacteria evolved to digest milk so the milk evolved to feed them. In the case of humans the reason why we produce such a complex range of HMOs must be to specifically advantage *B. infantis* over other bacteria.

So what do we get from the deal? Why do we want a single bug flourishing in our infant gut? The primary reason is probably competition. Human infants are born with a gut that is ostensibly sterile and provides an amazing opportunity for bacteria. It is warm, wet and full of a steady supply of nutrition from food and milk. It is also relatively unprotected by the naive infant immune system. From the moment the mother's waters break, the baby starts swallowing a range of bacteria. The vagina is full of a carefully controlled mix of bacteria. And what strikes anyone watching a normal vaginal delivery is that the baby is born, face down, into a pile of faeces. It was my job as a medical student to hold a little gauze over the stool, protecting the child and to some extent the dignity of the mother (today, with our advancing knowledge of the microbiome, I wonder if it would have been better not to). Part of the role of breast milk then is to encourage the growth of 'good' bugs. This is the most obvious way in which *B. infantis* protects us: by binding to the cells lining the baby's large intestine, preventing other, more harmful bacteria doing so. And crucially it seems to bind more strongly when grown on breast milk.

It's not just a matter of outcompeting the pathogens – *B. infantis* may keep them at bay with secretions. When grown on HMOs it produces short chain fatty acids (SCFAs). These are molecules that have become famous from their beneficial effects shown in adults eating a high-fibre diet, but they seem to be equally important in children. Some of them may directly kill harmful bacteria. Bruce describes this in terms of the concept of a 'shelf stable baby'; preserved from the inside by the secretions of the friendly bacteria. Other SCFAs like acetate may feed the developing infant brain. This role in brain development may in part explain the huge complexity of the HMO in human milk compared even to our closest chimpanzee cousins.

There is a catalogue of other benefits too. It's been shown to be anti-inflammatory, specifically in infant gut tissue when compared to adults. Understanding of the development of the infant immune system is still in its early stages, but Bruce thinks that the specific culturing of *B. infantis*, and the relative lack of diversity in the infant gut, may be crucial to the development of a mature immune system. In fact he is evangelical about this. He believes that the current epidemic of asthma, allergy and atopic disease in the US may be largely due to the loss of *B. infantis* from infant guts.

*B. infantis* also seems to reduce intestinal permeability, tightening up the joins between cells. Leaking infant guts may cause sickness directly but also affect the long-term development of the immune system. Evidence for this comes from an extraordinary study in Bangladesh. The dominant species in the stools of

A premature baby on the neonatal intensive care unit at Bordeaux Hospital, France. The profusion of tubes and wires enables continuous monitoring of physiological parameters and ventilatory support.

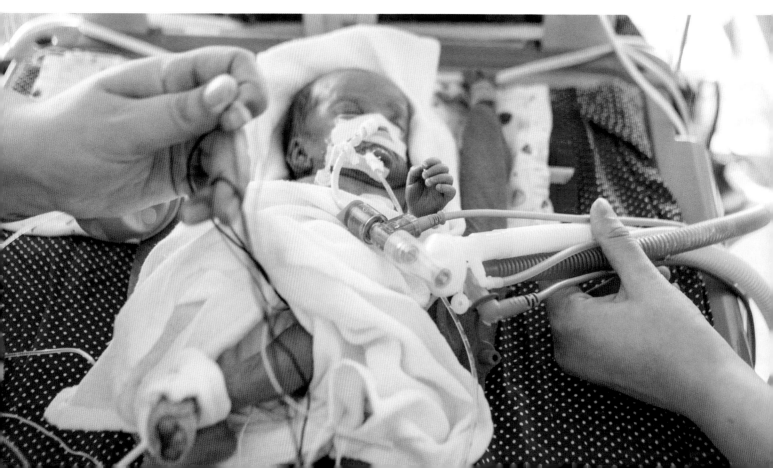

■

AT THE END OF MY CONVERSATION WITH MARK
I ASKED IF HE HIMSELF HAD BEEN BREASTFED.
'NO,' HE REPLIED. 'PERHAPS IF I HAD BEEN
BREASTFED I'D HAVE BEEN A SURGEON.'

■

the infants in the study was *B. infantis* and they were 96 per cent breastfed. The more *B. infantis* in the stool, the more weight gain and the better the responses to oral polio, tuberculosis and tetanus vaccines.

And there are the extraordinary little considerations that *B. infantis* makes in order to be a good houseguest. Other similar species of bacteria are not directly harmful but they do digest human mucus. When they do this they accidentally produce sugars that are useable by dangerous bacteria. By contrast *B. infantis* leaves the complex sugars in human mucus intact, starving the pathogens.

Just as the early efforts to understand the link between breast feeding and infant health required collaborations between chemists, physicians and microbiologists to make remarkable progress, Bruce German has forged a multidisciplinary team at UC Davis. While Bruce was starting to dissect the chemistry of milk in the lab, neonatologist Mark Underwood, at the UC Davis children's hospital, was treating and studying children born long before 37 weeks' gestation. Over 10 per cent of children are born prematurely in the United States and they face a few singular challenges. Inadequate lung development is the most immediate problem, but in the weeks spent in intensive care after birth a devastating condition called necrotising enterocolitis, or NEC, claims many infant lives. In NEC the tissue of the gut becomes inflamed and dies, allowing the contents of the gut to leak into the abdomen causing massive infection. It affects approaching 10 per cent of infants who are born weighing less than 1,500 g and half of those affected will die. In the decades he has spent caring for premature babies, the death rates from NEC have not changed significantly, but some clues that *B. infantis* may help to reduce this death rate are starting to emerge. Breast milk improves outcomes in NEC and additionally, a lack of *Bifidobacteria* seems to increase risk.

Neonatal intensive care is a dangerous place to be. Paradoxically this may be because it is too clean, or perhaps clean in the wrong way. Antibiotics and continuous cleaning keep *Bifidobacteria* at bay, but disease-causing organisms flourish in even the most fastidiously clean units. A trial is just starting at UC Davis but the evidence from other studies shows that administering *B. infantis* as a probiotic (a dietary supplement containing live bacteria that promotes health benefits) together with human breast milk serving as a prebiotic (a dietary supplement to stimulate the growth or activity of commensal microbes), may help to further reduce the incidence of NEC.

Bruce is enthusiastic about the use of probiotics even in term infants and suggested that I give my own child some *B. infantis* prebiotic. I asked why simply breast feeding wouldn't be enough. 'Because *B. infantis* is extinct in much of the developed world. It's not a bacteria which acquires resistance easily and particularly in the USA the use of formula milk for multiple generations has simply starved it out of existence.'

Bruce has the infectious enthusiasm required for truly visionary science but I wondered if I was being seduced by his ideas too easily. The genetic case was certainly persuasive that *B. infantis* had co-evolved with breast milk to be the main colonist of the infant gut. Why else would it have the entire genetic toolkit to use molecules found only in human breast milk, molecules which humans were totally unable to digest? But I wanted a clinical perspective so I asked Mark Underwood for his view. Mark is no less visionary than Bruce but he has the sort of quiet, clinical caution that comes from being a doctor in a speciality where a lot of child patients die. He was no less enthusiastic than Bruce. He believes the evidence stacks up from all sides and he is about to start a trial of giving *B. infantis* as a probiotic in the neonatal ICU. From the second week of life, I have been giving Lyra once-daily *B. infantis* supplements sent by Bruce.

At the end of my conversation with Mark I asked if he himself had been breastfed. 'No,' he replied. 'Perhaps if I had been breastfed I'd have been a surgeon.' He was being ironic: surgeons may think of themselves as being at the top of the tree, but physicians like to joke amongst themselves that surgeons are mere technicians. But his answer contained an interesting truth. He was a healthy, successful person. It is true that the mother–infant pair are what Bruce calls a 'powerful Darwinian engine' driving extraordinary evolutionary change. Together they have co-opted another species as the world's most effective nanny, supporting brain development and the development of the immune system, as well as fighting pathogens. And loss of *B. infantis* from our ecosystem may well explain the rise in allergic and atopic disease. But contained in Mark's answer is the idea that, despite multiple generations of formula-feeding and antibiotics rendering this seemingly vital bacteria functionally extinct, it's possible to become a healthy professor without it. People continue to live longer and longer. The human body has extraordinary resilience and redundancy; regaining *B. infantis* in our infant guts may well have wide-ranging benefits, but is testament to our adaptability that we can survive without it.

# GROWING UP, UP, UP …

When it comes to growing up, the van Kleef-Bolton family from London are world-class. At 6 ft 5 in (195 cm) and 7 ft (213 cm) respectively, Keisha and Wilco are the tallest couple in Great Britain and have only just been knocked off the global top spot, outranked by the lofty Chinese couple Sun Ming, 7 ft 8 in (233 cm), and Xu Yan, 6 ft 1 in (185 cm), in 2016.

While Keisha and Wilco are outliers at the far end of the distribution of human height, collectively as a species we have all gone through an incredible growth spurt in the last 150 years or so. Since the middle of the nineteenth century records show that the average height in industrialised countries has increased by about 10 cm. That's a serious increase in such a short space of time, and as far as we know is unprecedented. In fact the study of early human skeletons strongly suggests that human height stayed pretty much the same from the Stone Age until the mid-part of the nineteenth century. So what happened around the 1820s? Well, all of the available evidence suggests that this swift increase in height was not driven by any rapid-fire evolutionary selective pressures. The time frame is far too short for evolution by natural selection to play out and there is no reason to

Average height UK Male

Average height UK Female

think that height has been under particular selective pressure in the last 100 years or so. The environmental influences on the other hand seem to track very tightly with the increase in height. We know that if a child is malnourished or suffering from disease at particularly critical moments in childhood, they will never reach their full potential adult height. But since boys stop growing around their late teens and girls in their mid-teens, proper nutrition before puberty is essential to fulfil genetic potential for height. Protein, calcium, vitamins D and A all have an effect on height, and deficiency in all of these nutrients in the early nineteenth century was commonplace. But starting around the mid-1800s the punishing lives of populations through the early years of the industrial revolution began to give way to more widespread benefits, including better sanitation, clean running water and improved nutrition. Slowly this allowed the populations of countries like the UK to start fulfilling the genetic potential of human height. The truth is (as many parents instinctively know ...) that eating up your greens and drinking your milk really will make you grow up strong and tall.

The most startling journey from low to high across this time has been made by the Dutch. For the data shows that the average nineteenth-century Dutchman was looking up enviously at almost all of their European neighbours, but then a dramatic climb fuelled by increased living standards has taken them slowly but surely to the top of the global height charts. Although not without a few blips, both World Wars triggered a reversal in the upward trend in many countries as the availability of resources tightened dramatically. In recent years widespread increase in height has slowed down, stopped or even reversed. This is the case in the United States where a lack of free health care, and a diet high in calories but low in nutrients, may be the major contributing factors. It is likely that the Dutch are approaching the maximum height their genes will allow. Supplementation with extra vitamins, calcium and protein beyond the recommended daily amounts will not increase gains (and in fact there are large studies showing that excess vitamin supplementation shortens life).

But tall people aren't just tall because they have eaten better as children. Human height is determined by both genetics and environment. Your genes are a hand of cards you are dealt. Your environment is the way you play them. It's a case of nature via nurture. The major environmental influence on height is nutrition, affected by both diet and disease. Around 80 per cent of the variation in height between people is determined by their genes, and around 20 per cent can be attributed to the environment, although these numbers vary with different populations around the world.

You can work out how much of your height has actually been influenced by your parents demanding you clear your plate, and how much was set in stone from

**EAT YOUR ~~GREENS~~ PROTEINS**

During childhood the most important food that influences your final height is protein. Meat, fish, eggs, nuts, legumes and dairy products are all good sources of protein (which is why it is a considerable nutritional challenge to bring children up on a vegan diet). Other minerals, in particular calcium, and vitamins A and D also have a direct influence on height. For this reason malnutrition during the key stages of childhood can have a direct and significant effect on growth. This means good nutrition is particularly important before and around the growth spurts of puberty. For girls this begins around 10 years old and continues until their mid-teens when maximum height is reached. For boys it's later, with maximum height not being reached until the late teens.

UK's tallest couple Wilco and Keisha van Kleef-Bolton pose for a photograph with their 11-days-old son Jonah, 4-year-old Eva and 6-year-old Lucas near their home in Dagenham, in 2012.

the moment of conception. The average height of a man in the UK is around 5 ft 9 in (175 cm). Take me for example. I'm 6 ft 1 in (185 cm) so I'm 10 cm taller than the average. Eight of those 10 cm are determined by my genes (my dad is a 6 ft 4 in [200 cm] Dutchman) and 2 cm by my diet (my mother is, in the words of P. G. Wodehouse, 'God's gift to the gastric juices'). I tower over Xand by a full centimetre simply because I listened to mum a bit more.

You don't need to be a population scientist to see that tall parents beget tall children by passing on genes for tallness.

In the case of Keisha and Wilco van Kleef-Bolton, this certainly seems to be playing out predictably. They are the proud parents of five children, Lucas, the oldest at 11, is already 5 ft 4 in (c. 163 cm) and towering over his classmates; Eva, 8, is the average height of an 11-year-old; and 4-year-old Jonah is standing shoulder-to-shoulder with boys twice his age. While it's still a little too soon to judge the newest arrivals to the family, early indications point them to the skies as well: Ezra, the tallest of the 1-year-old twins, is in the 91st percentile and Gabriel is not far behind.

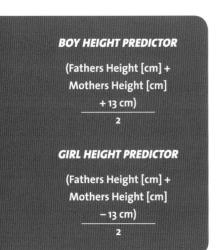

**BOY HEIGHT PREDICTOR**

$$\frac{(\text{Fathers Height [cm]} + \text{Mothers Height [cm]} + 13\text{ cm})}{2}$$

**GIRL HEIGHT PREDICTOR**

$$\frac{(\text{Fathers Height [cm]} + \text{Mothers Height [cm]} - 13\text{ cm})}{2}$$

But we don't really need to wait to see roughly how tall any of their children will be. Since the 1970s we've been using a rough and ready formula to predict the eventual height of offspring with nothing more than just the parents measurements. By simply adding the height of two parents together, adding 13 cm to the sum of the two numbers for boys and subtracting 13 cm for girls and then dividing the result by two (see left), you end up with a pretty good estimation for the height of the children. So in the case of the van Kleef-Boltons, the boys would be expected to be 210.5 cm, and the girls 195.5 cm.

This is not of course a precise calculation, but its rough reliability does indicate that height is a trait that is significantly inherited. That doesn't mean there is a single gene for height; very few traits have a direct one-to-one relationship. Instead your height, like many other characteristics, is controlled by a multitude of genes interacting with a multitude of environmental factors.

We now know that in the case of height your genes are about 80 per cent of the story in determining how tall you and your children will be. The reason we know this with such accuracy is because there have been a wide variety of studies that have explored the heritability of human height using a long-established method of teasing out the influence of nature vs. nurture.

The principle of these studies is simple. Take a group of identical or monozygotic twins, to use the technical term, like Xand and myself, twins who have developed from a single fertilised egg and so share 100 per cent of the same genes. You then compare a trait such as height difference between each of the identical twins in the group with a group of dizygotic twins, or non-identical twins (or even just siblings) who all share only about 50 per cent of their genes.

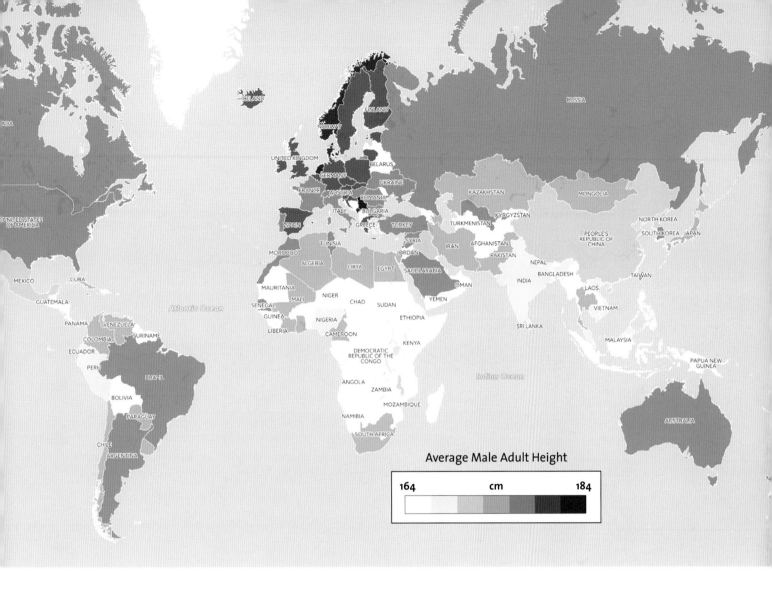

## Average Male Adult Height

164      cm      184

It is assumed that identical and non-identical twins grow up in equally similar or different environments, so this method of comparing groups of identical and non-identical twins enables you relatively easily to quantify the hereditability of a trait. So, for example, in the case of a height study, if it shows that the identical twins are considerably closer in height than the non-identical twins then this strongly indicates that genes play an important role. The actual analysis that can be applied to a study like this, both statistically and genetically is far reaching and complex but the principle remains the same – the greater the similarity between identical twins compared to non-identical twins the greater the heritability of the trait.

One of the most recent of these large studies conducted by Peter Visscher of the Queensland Institute of Medical Research in Australia looked at 3,375 pairs of Australian twins and siblings and found that the heritability of height is around 80 per cent. Other studies have come up with similar findings, including one that looked at 8,798 pairs of Finnish twins, in which the heritability was found to be 78 per cent for men and 75 per cent for women. Interestingly, similar studies in Asia and Africa have found the per cent heritability to be around 65 per cent or lower, because these regions tend to have populations that are less

Map showing variation in average adult male height in various nations across the world.

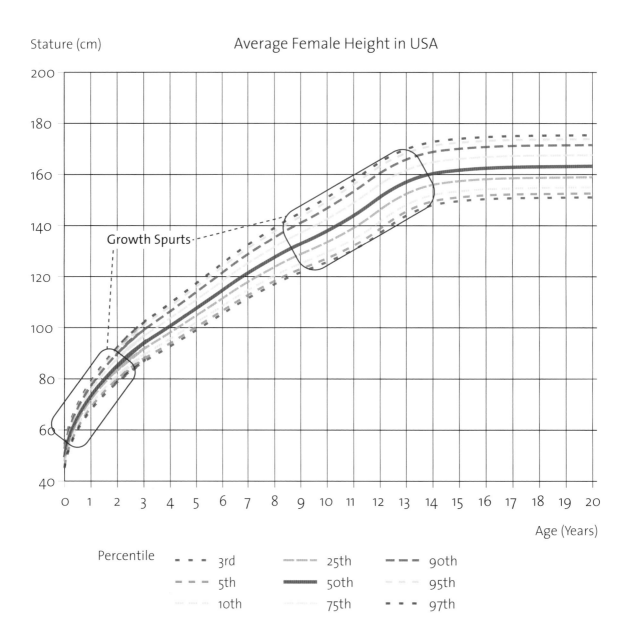

## Average Female Height in USA

Stature (cm)

Age (Years)

Growth Spurts

Percentile
- - - 3rd
- - - 5th
- - - 10th
----- 25th
▬▬▬ 50th
----- 75th
- - - 90th
- - - 95th
- - - 97th

Average female growth chart from birth to 20 years old: showing from the 3rd to the 97th percentiles.

mobile and so more ethnically and genetically defined compared to the greater genetic homogeneity we see in the west.

Regardless of the height you reach as an adult, the journey to get there is not steady, and only now are we beginning to truly understand the extraordinary process behind this rapid growth, a process that is dependent on an intricate interplay between your genes, brain, a cascade of chemicals and every bone in your body.

# GROWING PAINS

In the first six months of your life, you grew more than at any other time since. It's a growth spurt unlike anything else our bodies experience, with most of us growing a massive 30 cm in that first year. As a new parent this is particularly evident. It seems some days as if my daughter is growing in front of me. If we continued to grow at this rate, we would be 10 feet (~3 m) tall by the time we were 10 years old, but by the end of that first year that frenzied growth rate has slowed down and will continue at a far more subtle speed until the madness begins again at puberty.

The secrets behind the process of growth reveal how the body works as an integrated system, not just separate organs and limbs functioning in isolation. Starting at the business end of the process, the bones that really define your growth are the long bones of your body, in particular the femurs in your thighs, the fibulas and tibias in your lower legs, and the humerus, ulna and radius in your arms. These are the site of the major longitudinal growth during that first year of development. These bones don't just uniformly increase in size as they lengthen; the growth is focused around a particular part of the bone called the metaphysis found at the end of each long bone. If you looked at an X-ray of the metaphysis region of the fibula and tibia of a 10-year-old, you might conclude that the child has a broken leg. But what you are actually seeing in the 'fracture' across the bone is the location of growth, a line that is called the epiphyseal plate or growth plate. This is a soft disc made of hyaline cartilage (the same cartilage that you can feel in your nose) and it's here that cells called chondrocytes divide throughout the first 15 or so years of life and the rate of division increases furiously during a growth spurt. As the chondrocytes divide they secrete cartilage, a protein matrix that forms the template for bone, and the continuous division pushes the older cells towards the shaft. These gradually die and become 'mineralised'. The chondrocytes die, and cells called osteoblasts move in and secrete bone tissue into the cartilage. It's this process that results in the elongation of the bone – this is how we grow.

The long bones of the human skeleton.

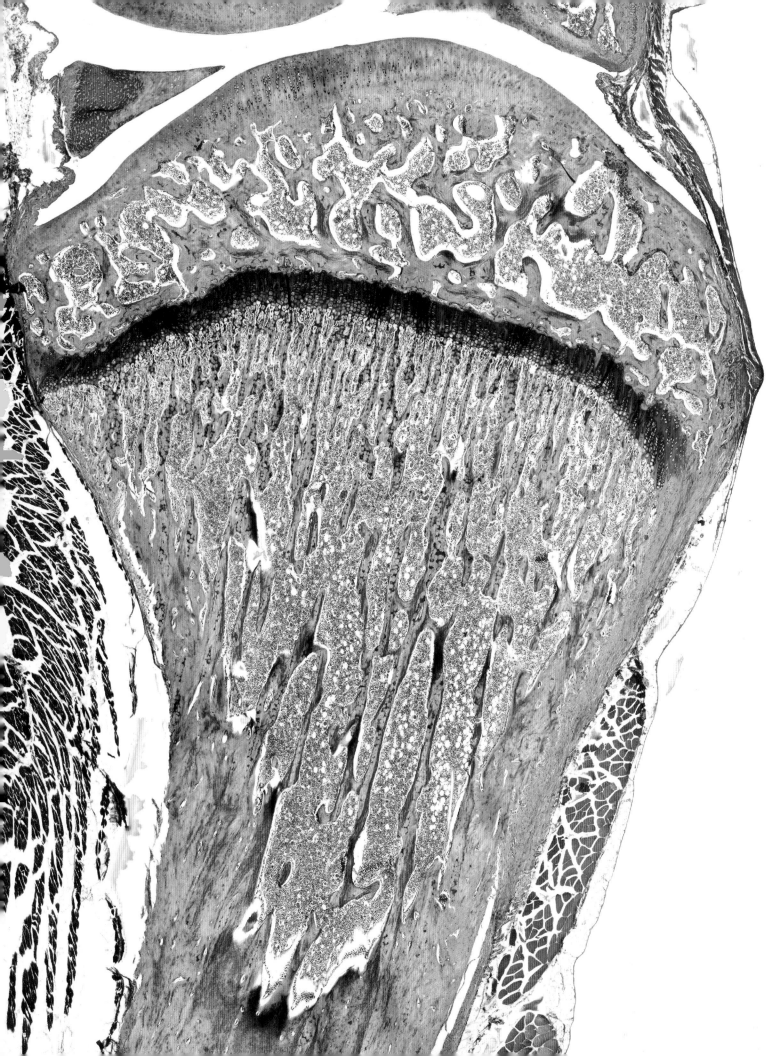

It's only once you reach adulthood that the activity in this area stops due to a process of programmed cell death (oddly controlled by oestrogen, the female hormone, in both boys and girls) and the growth plate closes and stops growing. The old growth plate becomes visible on X-rays as an epiphyseal line, a faint scar notched into your bones that you will carry for the rest of your life. At this point, bones can no longer elongate, growing any taller is now impossible.

## BRAIN–BODY INTERFACE

The full story of your miraculously extending bones starts far away from your skeleton. Nestled deep inside the centre of your brain, just behind your eyes, is a structure no bigger than the size of an almond, called the hypothalamus. It is from here that growth is controlled.

Your brain facilitates your conscious desires by sending signals to your muscles. This is your 'somatic' nervous system, the one that allows you to consciously move about, to speak, to look at things. But, in parallel, you have another subconscious, or autonomic, nervous system governed largely by the hypothalamus. It integrates more data than it's possible to calculate, from all your sense organs, your memory and experience, your cerebral cortex and amygdala, and it uses this data to control functions of your body that you likely take for granted. Digestion, heart rate, sweating, the size of your pupils and also growing. The hypothalamus is the link between the brain and the body.

Part of this regulation is control of the body's hormones or endocrine system. The hypothalamus secretes hormones itself which include vasopressin (which controls thirst and water reabsorption by the kidneys) and oxytocin (the 'love' hormone, which has a range of effects including stimulation of milk secretion and uterine contractions).

But most of your endocrine or hormone system is located around your body in specialist endocrine organs, like your thyroid, gonads or adrenal glands. The hypothalamus controls these organs remotely through a cascade of hormone signals sent first to the pituitary. The pituitary gland is around the size of a pea and dangles beneath the hypothalamus from the underside of your brain, on a stalk. It sits behind and between your eyes resting in a little bowl of bone in the base of your skull called the sella turcica or Turkish seat. Via this tiny organ your hypothalamus controls your reproduction, sex drive, lactation, metabolism and of course your growth.

**Opposite page:** Light microscopy of the end of a growing humerus. The dark blue staining band is the epiphyseal growth plate consisting of dividing, maturing and dying chondrocytes. The cartilage made by the plate serves as a temporary substrate on which immature bone will be deposited and before becoming mineralised adult bone. Above the growth plate is the epiphysis, covered with curved articular cartilage. Below the plate is the metaphysis, the flared region of bone extending into the shaft or diaphysis. The cavities in the bone contain the bone marrow where blood is made.

Roger Moore, the best James Bond, fights with Richard Kiel (7 ft 1.5 in), as 'Jaws', in a scene from the film *The Spy Who Loved Me*, 1977. Kiel's height and features were a result of excess of growth hormone secretion before and after the growth plates had closed. Once growth plates are closed height can no longer increase but the voice deepens, and the brow and hands thicken in a condition called acromegaly.

It's not a simple process. The hypothalamus secretes the unimaginatively named growth hormone releasing hormone (GHRH) or growth hormone release inhibiting hormone (GHRIH). These in turn signal to the pituitary to release or stop releasing growth hormone. Growth hormone then directly acts on the cells of your body instructing them to divide and stimulates the liver to produce insulin-like growth factor 1 (IGF-1) which also makes you grow. It takes the brakes off cell division and causes the growth of almost every cell in the body.

Levels of IGF-1 and growth hormone can be affected by a huge range of processes which feed into the hypothalamus: insulin levels, disease, protein intake, stress, genes, physical fitness and sex hormones.

We see this kind of signalling cascade with almost all biological processes, whether it's the immune signalling pathways inside cells that Chris studied in his

PhD, or the whole body cascades of chemical signals from organ to organ. They allow for delicate control of biological processes at multiple levels, with each organ feeding back information to regulate the process. They are also remnants of our evolutionary past. As organisms became more complex, it was easier for evolution to add another layer of control than to redesign from scratch. As we've seen in other chapters, we still have ancient systems in our modern bodies but with extra lines of code to allow for more regulation.

We all produce growth hormone (and thus IGF-1), every day throughout our lives. In adulthood the average healthy individual produces about 400 micrograms

Dutch Giant Jan Van Albert with his wife and Dwarf Beppetoni. 1911.

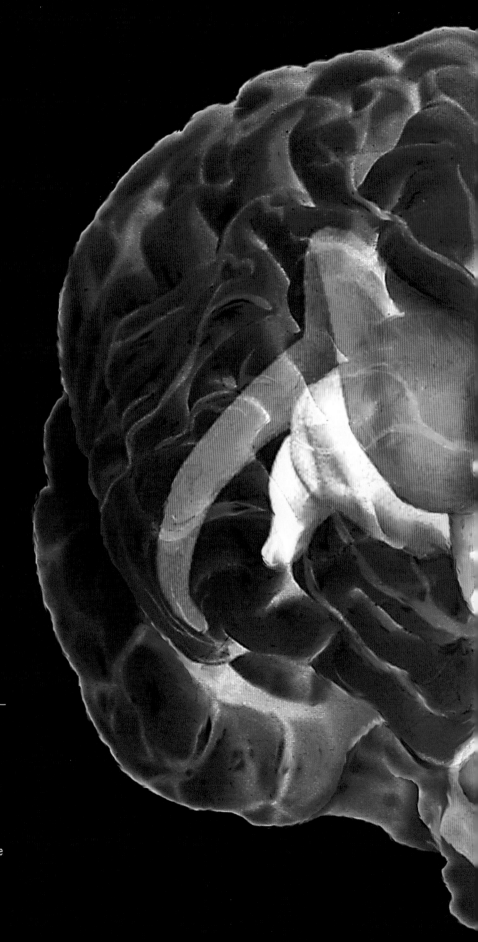

A coloured composite 3-D magnetic resonance imaging and computed tomography scan of a human brain, seen from the front. The ventricles (pink) circulate the cerebrospinal fluid, which cushions the brain. Beneath the lateral ventricles lie the sensory-processing thalami (orange) and the hypothalamus (green, centre), which controls emotion and body temperature and releases chemicals that regulate hormone release from the pituitary gland (green, above fourth ventricle, which is pink).

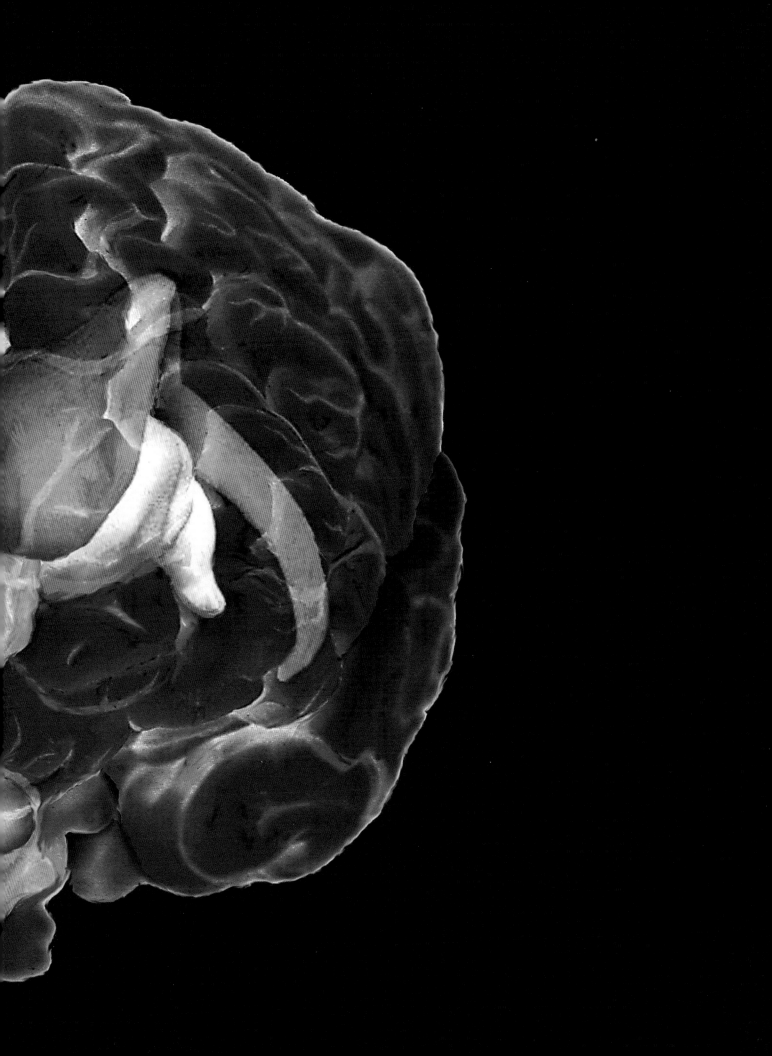

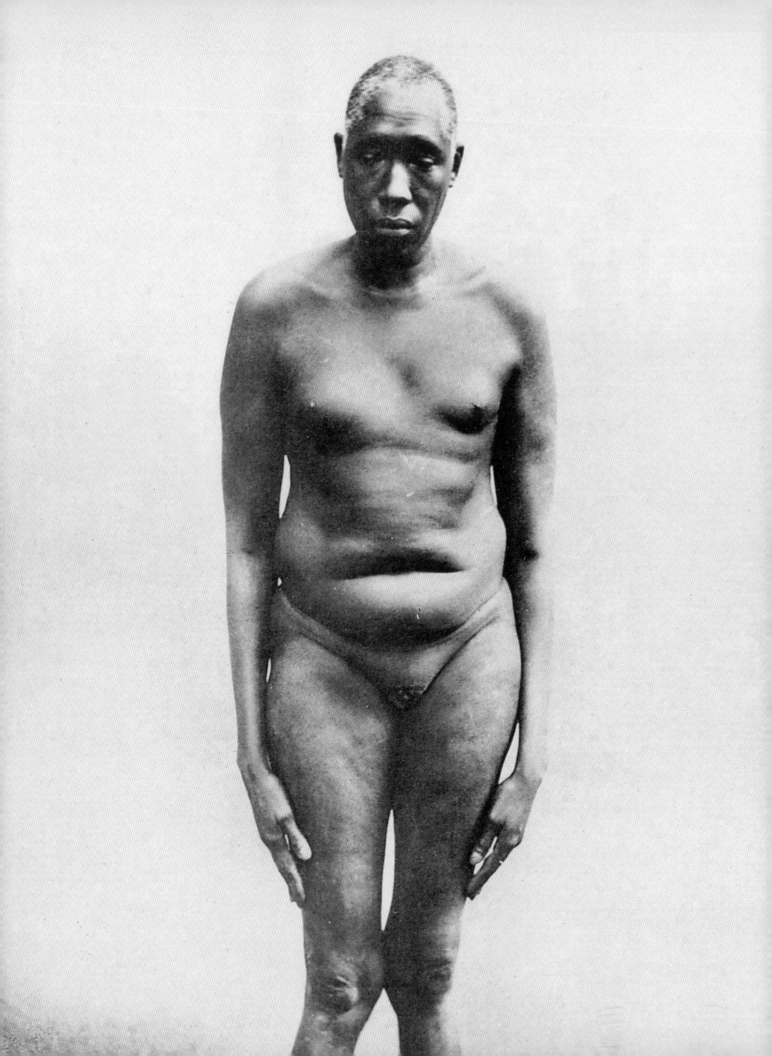

a day (a scarcely visible amount), and it plays a crucial role in the maintenance and renewal of our bodies as well as controlling a host of other bodily functions. In children and teenagers, the levels of growth hormone are much higher, reaching 700 micrograms a day in the midst of our most rapid periods of growth, and it's these levels that drive the process in the growth plates of young bones. In this way, through the cascade of hormones from the brain, which travel through the blood vessels of your body to command the cells in the growth plates of your long bones to divide and push those bones a little longer, millimetre by millimetre, you grow.

As is often the case in medicine, we have understood how body systems function in healthy people by studying those for whom these systems have gone wrong. Dwarfism occurs when IGF-1 is not produced or when the receptors on the cells' surfaces that should detect it are absent or defective.

Conversely, tumours of the pituitary may secrete excess growth hormone, and if this happens in childhood prior to fusion of the epiphyseal plates, then gigantism results. Although these tumours are extremely rare in childhood, they have

**Opposite page:** A man with hypogonadism or eunuchism. This can be caused by pituitary dysfunction or a primary problem with the testicles. Note the small genitals and lack of secondary sexual characteristics due to lack of testosterone.

**Below:** A baby being weighed a week after birth. This child lost 7 per cent of her body weight, as is common during the first week of rapid growth. This weight loss is largely fluid and some fat lost whilst feeding and digestion catch up with growth.

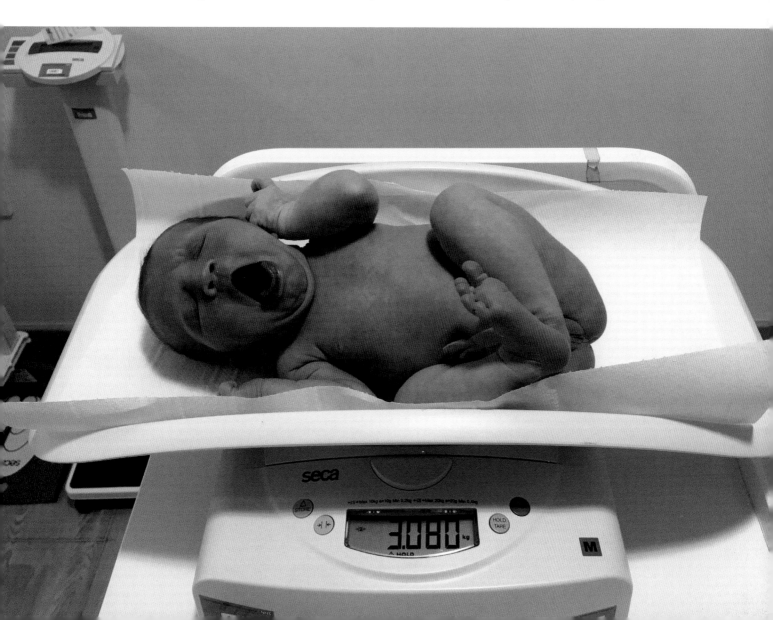

produced two extremely well-known actors, including Richard Kiel (the infamous 'Jaws' villain from two James Bond movies) and Andre the Giant, a wrestler and actor from *The Princess Bride*.

As well as being giants over 7 ft (213 cm) tall, both of these stars exhibited the other effects of excess growth hormone secretion, a condition called acromegaly. If growth hormone secretion occurs after the long bones have fused, then you can't grow any taller, but bones and other tissues continue to grow. The brow ridge and jaw thicken, the tongue and hands become vast and thick, and the voice deepens. In athletes using growth hormone as an illegal performance-enhancing drug, jaw changes often necessitate orthodontic braces to realign teeth – a subtle tell-tale sign for doping.

It's a beautiful, complex cascade that we have been able to understand in greater and greater detail through the revolution in molecular biology and genetics over the last 50 years, but one particularly strange thing about our growth through childhood and adolescence has remained a mystery. Unlike any another primate, we have a very odd pattern of growth through to adulthood. As we've already seen in this chapter, the first six months of life witness the most rapid period of growth, but then this slows dramatically through the next 10 years – a time we humans call childhood. Unlike any of our nearest relatives, including chimpanzees and bonobos, we grow at a fraction of the maximal rate through this period. It's as if the race to adulthood is on hold, until suddenly we burst into activity again around the age of 10 as we experience the growth spurt of puberty. The mystery is why. Why do we all follow this oddly stunted pattern of growth? In the last few years an intriguing hypothesis has emerged to explain the biological oddity we call childhood.

# GOOD THINGS COME TO THOSE WHO GROW

In June 1765 Daines Barrington, the British lawyer, naturalist and distinguished fellow of the Royal Society, made his way the one mile from his home in King's Bench Walk in the heart of legal London, to a rather less respectable address on the east side of Soho. The reason for his journey into this more unsavoury area of London was to visit the temporary occupants of 21 Frith Street – an Austrian man named Leopold and his two children 14-year-old Nanneri and 8-year-old Wolfgang.

Under his arm Barrington carried a clutch of documents and papers, but most importantly a newly composed music manuscript written in a 'challenging, contemporary Italian style'. The purpose of bringing the manuscript was to place

'Account of a Very Remarkable Young Musician. In a Letter from the Honourable Daines Barrington, F. R. S. to Mathew Maty, M. D. Sec. R. S.', Daines Barrington, *Philosophical Transactions*, 1 January 1770.

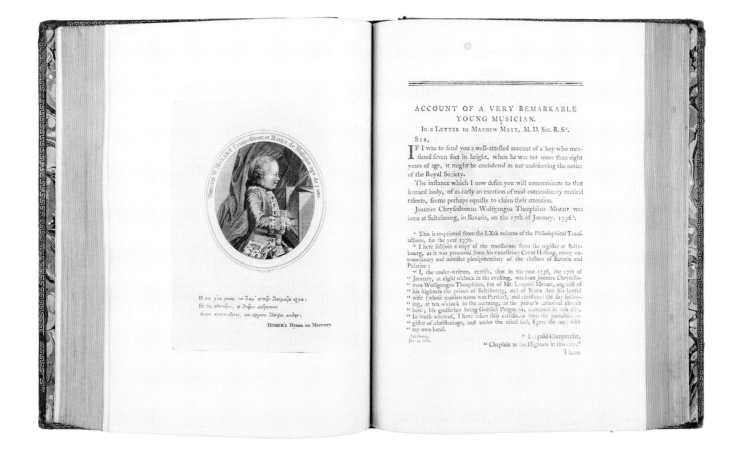

∎

'EVEN TODAY, AFTER A CENTURY OR SO OF
SCIENTIFIC STUDY OF CHILD DEVELOPMENT,
PRECOCIOUS TALENT REMAINS A MYSTERY. WE ARE
STILL AS CURIOUS ABOUT TALENT NOW AS PEOPLE
WERE IN THE EIGHTEENTH CENTURY.'

**PROFESSOR UTA FRITH**

∎

it in front of the young boy Wolfgang so that Barrington could check for himself whether the rumours that had spread across London regarding this boy's precocious musical talent were really true. The boy was of course Wolfgang Amadeus Mozart and Barrington's test would be an easy trial for him to pass. Just by sight, the young Mozart played the piece effortlessly and perfectly, at the very first time of trying. 'The score was no sooner put upon his desk than he began to play the symphony in a most masterly manner, as well as in the time and style which corresponded with the intention of the composer,' he wrote.

Barrington went on to further test the abilities of the 8-year-old boy, challenging him to improvise a song of 'love and a song of rage'. Writing in a now famous letter to the *Philosophical Transactions of the Royal Society* some years later, Barrington described how Mozart's

> astonishing readiness, did not arise merely from great practice; he had a thorough knowledge of the fundamental principles of composition ... and his transitions from one key to another were excessively natural and judicious.

Barrington's visit, tests and subsequent publication of his observations are widely regarded as one of the first examples of Behavioural Science. As Professor Uta Frith, a current FRS and one of Britain's most distinguished cognitive scientists, wrote some 250 years later, 'Naturally, the methods of observation he used are rather crude to our modern eyes, but, the crucial point is that he gives concrete examples of behaviour and not just opinions.'

It did of course not just take Barrington to prove that Mozart was undoubtedly a child genius. The historical records are full of details of his precocious talent, from his first compositions as a 4-year-old, to his first symphony, composed during

that extended stay in London. This was a childhood that was truly full of extraordinary achievement, a unique talent that was maturing before the eyes and ears of the world. As Frith went on to conclude, 'Even today, after a century or so of scientific study of child development, precocious talent remains a mystery. We are still as curious about talent now as people were in the eighteenth century.'

Achievement and emerging talent however, is not something that is in short supply with children of 4, 5, or 6 years of age. This is the moment that many of us sit our children down at the piano for the first time, sign them up for the local football team or send them off to ballet class as well as seeing them grasp the fundamentals of reading, writing and arithmetic skills that they will carry throughout their lives.

Subconsciously or not, we are aware that this is a precious time, a moment when children are more than just sponges; they are receptive to developing new skills and abilities with an ease that will not be repeated at any other time in their lives.

Geoffrey Tozer (1954–2009) was an Australian classical pianist and composer, here performing at the age of 15, in July 1970. A child prodigy, he composed an opera at the age of 8, and became the youngest recipient of a Churchill Fellowship award aged 13.

The foundation of all of this new-found knowledge, skill and ability is of course the brain, and intriguingly we now think the brain power that goes into all of this intensive learning is intricately linked to that mysterious and odd pattern of growth we were puzzling on earlier in the chapter.

On a daily basis your brain demands a huge amount of the energy your body uses. Weighing around 1.4 kg, just 2 per cent of our total body weight, the average adult human brain consumes 20 per cent of our body's energy expenditure (to be precise that is 20 per cent of the resting metabolic rate [RMR]). To put this massive power demand into some context, if your body needs 1,400 calories just to sit on the couch all day doing sod all (that's what the RMR is), then your brain will be consuming 280 of those calories just to keep things ticking over, like deciding which channel to watch, or when to eat dinner. Put another way, it takes one Mars bar plus an extra bite for your brain to exist. No other organ in the body is so hungry for energy, but what is interesting is that the energy demands of the brain are far from constant throughout your life.

In 2014 a team led by Chris Kuzawa from Northwestern University in Illinois published in *Proceedings of the National Academy of Sciences of the United States of America* an intriguing paper entitled 'Metabolic costs and evolutionary implications of human brain development'. The title may sound complex but the aim of the study and the conclusions it came to are beautifully simple and may well provide our current best explanation for the long, slow process of human childhood. In the study the group measured two key factors across a range of ages from birth to adulthood. The first was to measure the brain's use of glucose at every stage of childhood from 0–15 years of age. This data was collected from a range

'THE BRAIN'S METABOLIC REQUIREMENT PEAKS IN CHILDHOOD, WHEN IT USES GLUCOSE AT A RATE EQUIVALENT TO 66 PER CENT OF THE BODY'S RESTING METABOLISM AND 43 PER CENT OF THE BODY'S DAILY ENERGY REQUIREMENT.'
**CHRIS KUZAWA**

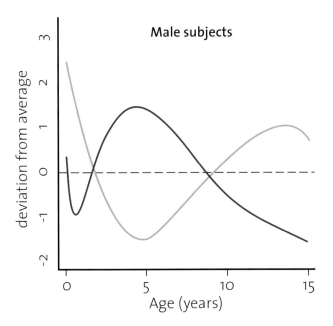
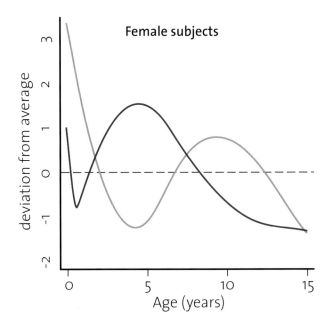

of previous PET and MRI studies and then collated into age order to produce the graph you can see above. The second set of data was an even simpler task involving the collation of available body weight growth across the same period of time. Individually these two graphs tell you very little, but put them on top of each other and an intriguing and powerful story about all of our lives begins to emerge from the data. It seems according to the findings of this study, that the more energy your brain demands the less your body grows. It's as if there is a switch from the body taking preference in the first year or so, to your brain taking preference in the childhood years, until the body takes over again at or just before the onset of puberty. It seems the secret story of our strange, long childhood is played out in the energetics of it all.

As Kuzawa and his team concluded:

> We find that the brain's metabolic requirement peaks in childhood, when it uses glucose at a rate equivalent to 66 per cent of the body's resting metabolism and 43 per cent of the body's daily energy requirement, and that brain glucose demand relates inversely to body growth from infancy to puberty. Our findings support the hypothesis that the unusually high costs of human brain development require a compensatory slowing of childhood body growth.

In other words the physical process of growing up is put on hold for almost 10 years whilst our brains explore all the riches the world has to offer. Childhood

Results from Chris Kuzawa's research: the red curves show the rate of glucose uptake by the brain; blue curves, rate of body weight growth over the same period.

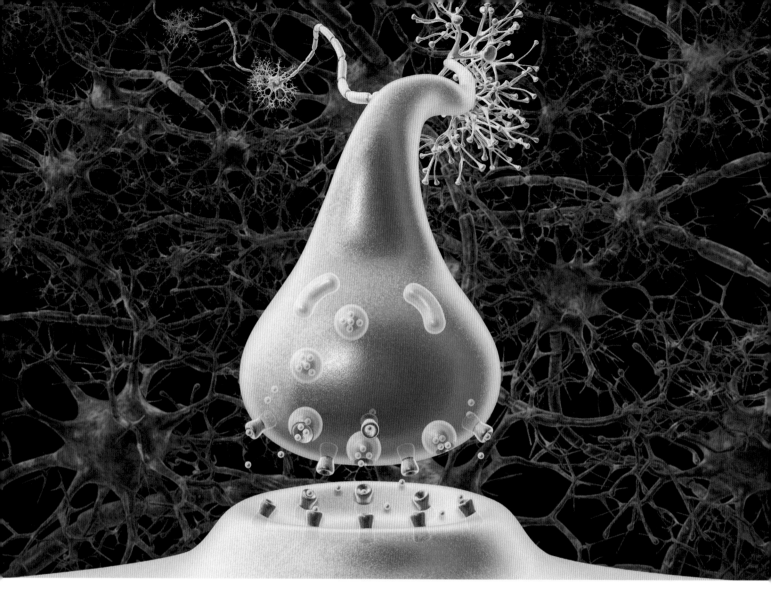

Illustration showing the action of neurotransmitters such as serotonin and noradrenaline in the synaptic cleft. Vesicles containing the neurotransmitter (green) move towards the pre-synaptic membrane where they fuse with the cell membrane, releasing their contents into the synaptic cleft. The neurotransmitter molecules act on the post-synaptic cell by binding to specific receptors on the cell surface (purple). They can also be taken back up by the pre-synaptic cell via other receptors (orange) for re-use.

is uniquely human, because the growth of our brains is unique – to survive we need to do an extraordinary amount of learning.

So if childhood is essentially (in its most reduced form) a process of energy diversion to the brain, then what is all the energy actually for? To answer this question we have to take a microscopic look inside the cells of the brain.

Much of the activity in the brain during childhood is about making new connections. We are born with a brain full of 80 billion or so neurons, but it's the way these neurons link together that really matters. Neurons connect at junctions called synapses. The electrical signals travelling down one neuron cause the release of chemicals, neurotransmitters, across the synapse. When the neurotransmitters bind to the next neuron they can either stimulate it, or stop it signalling. Synapses turn individual brain cells into a network that can exchange, store and process information. As we will see in the 'Learn' chapter, the creation and strengthening of new synapses is a key part of learning new skills and laying down new memories.

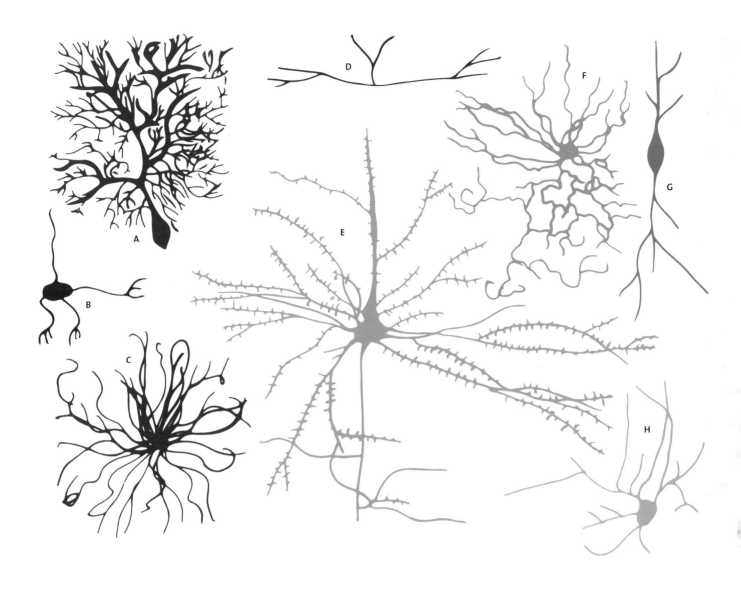

Different types of neurons. A. Purkinje cell; B. Granule cell; C. Motor neuron; D. Tripolar neuron; E. Pyramidal cell; F. Chandelier cell; G. Spindle neuron; H. Stellate cell. Based on reconstructions and drawings by Santiago Ramón y Cajal.

Synapses start forming long before you are born, but it's only after birth as the world rushes in through your senses that the rate of synapse formation really picks up. In the first three years of life, synapse formation is at its fastest. It is estimated that you form up to 40,000 every second, and by the time you head beyond your third birthday, it's estimated that you have a staggering one quadrillion ($10^{15}$) connections in your brain. That means on average every single neuron has around 7,000 synaptic connections. But then something intriguing starts to happen: rather than continuing to go up, the number of synapses begins to reduce, slowly but surely decreasing with age until by adulthood it's estimated that you have between 100 and 500 trillion ($10^{14}$ – $5 \times 10^{14}$).

The activity inside a child's brain may seem contradictory. On the one hand, by the age of 5 the metabolic rate of our brains is in overdrive (nearly twice that of an adult), suppressing the growth rate of our bodies which has slowed to a minimum, but at the same time, the number of connections within the brain is reducing by

Opposite page: Astrocyte brain cells (also known as astroglia) are a type of neuroglial cell that provides support functions in the central nervous tissue and spinal cord. They perform a variety of tasks, from axon guidance and synaptic support, to the control of the blood—brain barrier and blood flow. Immunofluorescent light micrograph of brain cells from the cortex of a mammalian brain. The nucleus of each cell is stained blue and cytoplasm stained green. Astrocytes may also play a role in information storage. The blue nuclei of other cells are also seen. Immunofluorescence is a staining technique which uses antibodies to attach fluorescent dyes to specific tissues and molecules in the cell.

millions of synapses each day. So what is happening? Why is so much energy being diverted into an organ that is appearing to deconstruct itself?

Well, at this period of our lives we are still making new synapses, but this synapse growth is random – a mind-boggling amount are made each day but not all are ultimately useful. So the brain keeps the useful synapses, those that are being constantly employed processing the new skills and experiences of childhood, but if a synapse isn't utilised it will eventually be snipped away in a process called synaptic pruning. A process that we are only now realising is crucial to the developmental growth of the brain all the way through childhood and up to adulthood.

To understand the process of synaptic pruning, we need to explore the other cells that make up your brain beyond just the neurons. As we have already seen your brain contains around 80–100 billion neurons but there is another type of brain cell that is far more abundant than this. Glial cells outnumber neurons by at least ten to one. Long thought of as the support act to the neuron's starring role it's only in recent years that we've begun to understand the extraordinary part these cells play in the development and function of the brain. Far from just support cells, they take on a wide range of functions in the brain, from creating and circulating the cerebral spinal fluid that the brain is bathed in, to controlling the delicate chemical balance around the neurons, to producing the myelin that is so crucial to the brain's function. But the glial cells we are interested in have an even more intriguing function. Microglia are the brain's first line of defence, the resident immune cells of the brain primed to fight off any invading bacteria or viruses. As macrophage cells they are, just like their brethren in the rest of the body's immune system, primed to identify and destroy anything that is recognised as foreign to the brain. From pathogens, to damaged or disfunctioning brain cells, to cancer cells, the microglia are there to engage, engulf and destroy anything that is seen as a threat to the brain.

Because the microglial cells were seen as a defensive system in the brain, the assumption was that when the brain was not under attack the microglia sat silently ready and waiting to jump to its defence. But in recent years that assumption has changed dramatically: we now know that microglia are active way beyond disease, a secret weapon in the development and function of the brain, and a critical component in the process of learning.

Rather than sitting idly waiting for the next virus, bacteria or knock on the head to kick them into action, the microglia are in fact the fastest-moving structures in the healthy brain, patrolling every area not just looking for disease but another equally destructive threat – inactivity. These are the cells that perform

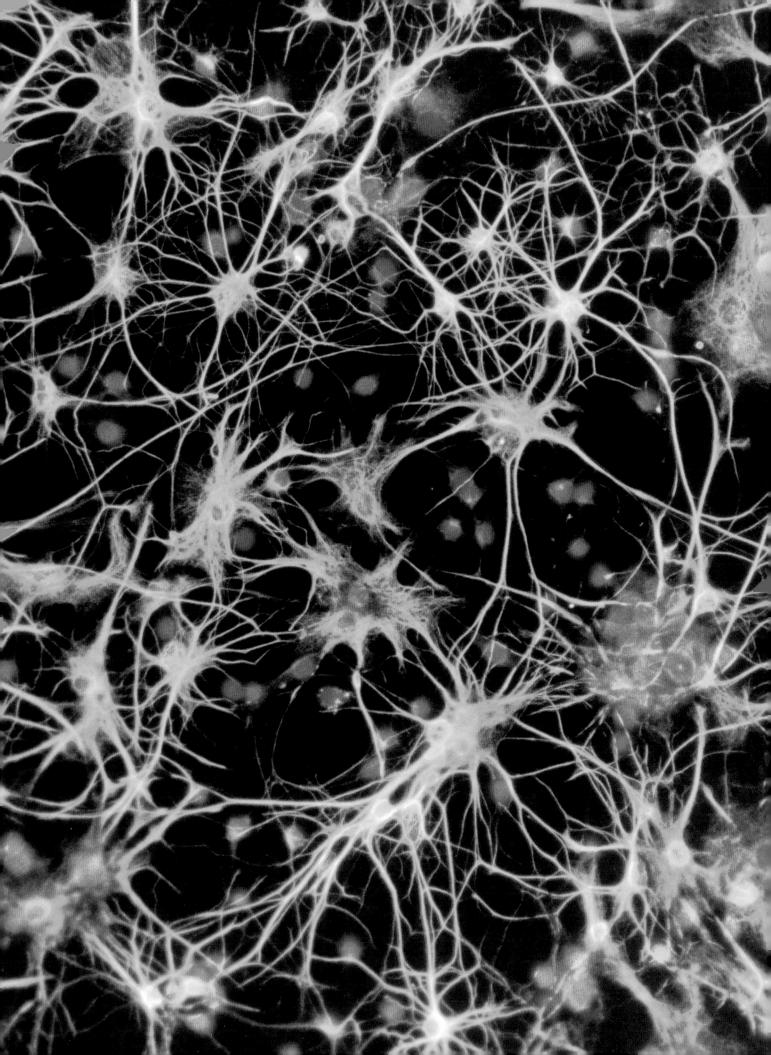

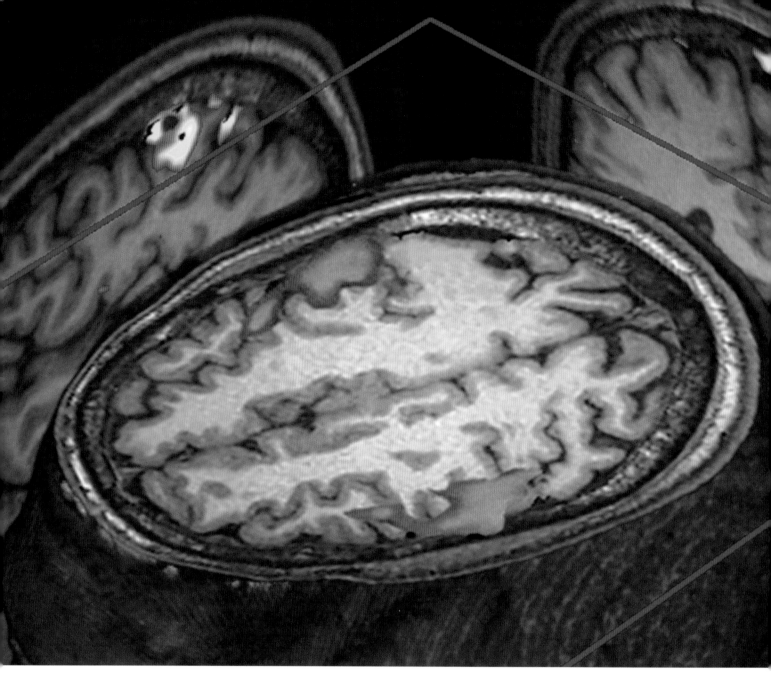

A subject, lying in the fMRI scanner, listens to a piece of instrumental music. The image shows brain areas responding to a meaningful auditory stimulus (beyond the sounds of the MRI apparatus).

the synaptic pruning that transforms the brain through childhood, searching out and destroying synapses that are inactive or underused. In this way the brain is continually sculpted by the microglia limiting redundancy, a process that we think is crucial to the learning and development we see through childhood. We are now also beginning to understand how, crucially, the microglia differentiate between those synapses to keep and those to destroy. Recent studies are revealing that a signalling system exists in the brain, with synapses that are needed given a chemical tag that says: don't attack me – whilst less active synapses are left defenceless. This microglia system is so critical to the healthy function and development of the brain that we now think it could be involved in the development of many diseases, including Alzheimer's, schizophrenia and conditions such as autism.

So here it seems is the answer to our conundrum: throughout childhood the developing brain demands huge amounts of energy, so much so that the growth of the body is put on hold. Inside the brain this energy is put to use not just building new networks and pathways but continually monitoring and sculpting the brain, leaving in place faster, more efficient brain circuits tailored to your needs. This is how we learn, by building up and knocking down, until by 11 years old your brain is almost fully grown. You can test this out next time you meet a 12-year-old, as we often do while filming for Children's BBC. It is often striking that their reasoning, intelligence and humour feel very grown-up, in stark contrast to their physical appearance. Of course ... you have to catch them in the right mood. The formation of synapses and neurons is complete, the massive demand for energy slows and the body can now use its reserves for the next stage of our development – puberty.

## MOZART'S BRAIN – AN END NOTE

We will never know exactly what magic was going on inside the young Mozart's brain but, almost 230 years after his death, a recent study into the brains of elite musicians may have given us a small insight into Wolfgang Amadeus's brain. A 2016 study carried out by scientists at the Concordia University in Montreal used MRI scans to explore the brains of 36 professional adult pianists, some of whom started playing before the age of 7 and some after. Unsurprisingly all the elite pianists showed increased amounts of grey matter in specific areas of the brain. Grey matter is made of unmyelinated neurons. It forms the surface on the cortex of the brain, where 'thinking' takes place. The increases were partic-ularly in areas connected with learning and memory, compared to non-playing individuals. Intriguingly they also showed a reduction in grey matter in certain areas of the right brain in comparison to the non-musicians. The results suggest that areas involved in repetitive learning are built up by the demands of musical training, but at the same time areas involved in sensorimotor control, auditory processing and sight reading are reduced. If you don't use it, you lose it!

Even more intriguing was the fact that the pianists who started before the age of 7 were not only objectively technically better performers (particularly in the left hand!) but also had a smaller amount of grey matter than the late starters in a key region involved with motor control. This suggests greater efficiency in the early starters and perhaps points to the handiwork of the microglia in shaping the brains of these budding young Mozarts.

# THE POWER OF PUPPY FAT

Acne vulgaris on a 16-year-old boy. Acne is a common chronic skin disease where pilosebaceous units (hair follicles and their accompanying sebaceous gland) become blocked. It affects mostly the face, back and chest.

As we've seen throughout this chapter and will see throughout the book, the human body is gradually giving up many of its most tightly held secrets to our ever more detailed investigations. But one of the truly great pleasures of studying human biology is that, no matter how hard we try, some mysteries just won't budge – and what kick-starts our extraordinary changes in adolescence is one of them.

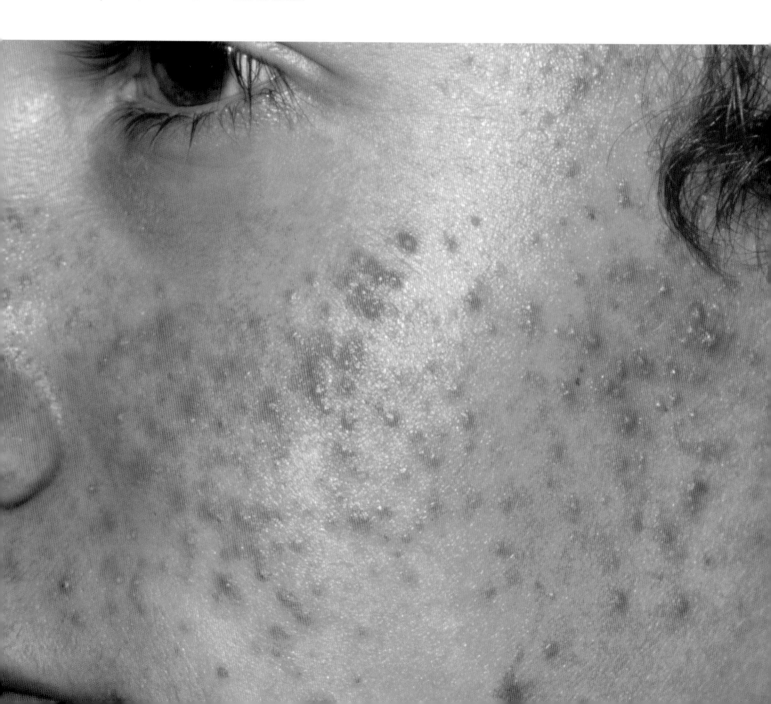

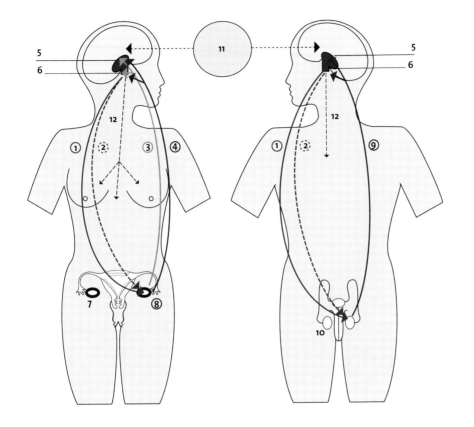

The body's homeostatic mechanisms are all controlled with cascades or pathways of signals. This allows for delicate control and feedback. The endocrine system is the paradigmatic example of this with signals going from the higher brain to the hypothalamus, then the pituitary then the endocrine organs which secrete hormones which often affect almost all tissues.

**Legend:** 1. Follicle stimulating hormone; 2. Luteinising hormone; 3. Progesterone; 4. Oestrogen; 5. Hypothalamus; 6. Pituitary gland; 7. Ovary; 8. Human chorionic gonadotropin; 9. Testosterone; 10. Testes; 11. Incentives; 12. Prolactin.

Puberty is the greatest change we go through in our lives, a wholesale metamorphosis that sees our bodies transformed from child to adult, the final hurdle in the journey from baby to baby-maker. I certainly remember it. I remember acne, body odour and the moment when the daft little moustache that grows on boys around the age of 12 needed to go. My shyness asking for Dad's razor. And above all, the immense sexual frustration of becoming 'reproductively competent' at a boys' school. Luckily Xand and I vented this frustration through work and sport. Throughout medical school and my PhD, I leant heavily on the knowledge acquired at high school. In retrospect, we got off pretty lightly, but I remember my own angst, and the angst of others almost as clearly.

The emotional turbulence is driven by the fact that there is no set pattern to puberty – ultimately its destination is predictable but each one of us takes a highly personal journey through it. Take any group of 14-year-old boys and girls and some will look and act like children, whilst others will already appear to be fully formed men and women. Xand and I were neither fast developers, nor slow. But I can remember the stark contrast between some of our contemporaries. One friend on the rowing team needed to shave twice a day while others retained childish

A STUDY OF RECORDS GOING BACK TO THE MID-NINETEENTH CENTURY SHOWS THAT BETWEEN 1840 AND 1950 THERE WAS A DROP OF FOUR MONTHS PER DECADE IN THE AVERAGE AGE OF MENARCHE IN WESTERN EUROPEAN GIRLS.

physiques and voices almost until we left to university. We almost all find ourselves in adolescent peer groups that contain a wild range of individuals at vastly different stages of progress, a dynamic that often fuels the volatility. Yet despite now having an incredibly detailed knowledge of the process and hormonal cascade that activates it throughout the body the reason for such variation is a mystery. We still don't fully understand what triggers puberty in individuals to start as young as 9 or as late as 16, with the major milestones like the first period for girls or ejaculation for boys distributed somewhere in between.

What we do know is that just like the growth spurt through adolescence, puberty all starts in the head. The hypothalamus and pituitary send out pulses of hormones that are carried through the bloodstream kick-starting the ovaries or testicles into action. In turn these gonadal tissues start flooding the body with sex hormones – testosterone for boys and oestrogen for girls but it's not the first time this system has kicked into action. In the first six months of life each one of us goes through a stage where exactly the same cascade of hormones occurs and levels of oestrogen or testosterone climb just as high as in the full throes of puberty.

This 'mini-puberty' is part of the rapid growth spurt we have already discussed in the chapter but it is also thought to be heavily involved in the 'genderisation' of the brain, influencing the gender-specific behaviours that define us through childhood and beyond. During the rest of childhood this hormonal system lies dormant, suppressed until it is reactivated 10 or so years later to drive us to sexual maturity.

It's surprising to realise that the control mechanism for such a monumental moment in human development remains stubbornly elusive and only in recent years has the mystery started to crack.

When the prestigious *Science* magazine published its 125th anniversary edition in June 2005 with '125 Big Questions' – 'What triggers puberty?' was one of those questions. Back then the answer was pretty much – we don't know – but now through progress in our understanding of genetics, epigenetics and the environmental factors that influence the biological mechanisms of puberty, we are beginning to piece together the story.

What we are finding is that there is almost certainly no single trigger for puberty. Like almost all complex processes in the body, it's a mixture of nature and nurture, inheritance and environment that ultimately work together to act as the trigger for our body's greatest events.

Genetics, of course, plays its part, with a number of twin studies suggesting genes account for around 45 per cent of the variation in the timings of puberty.

Puberty is a time of soaring highs and crushing lows.

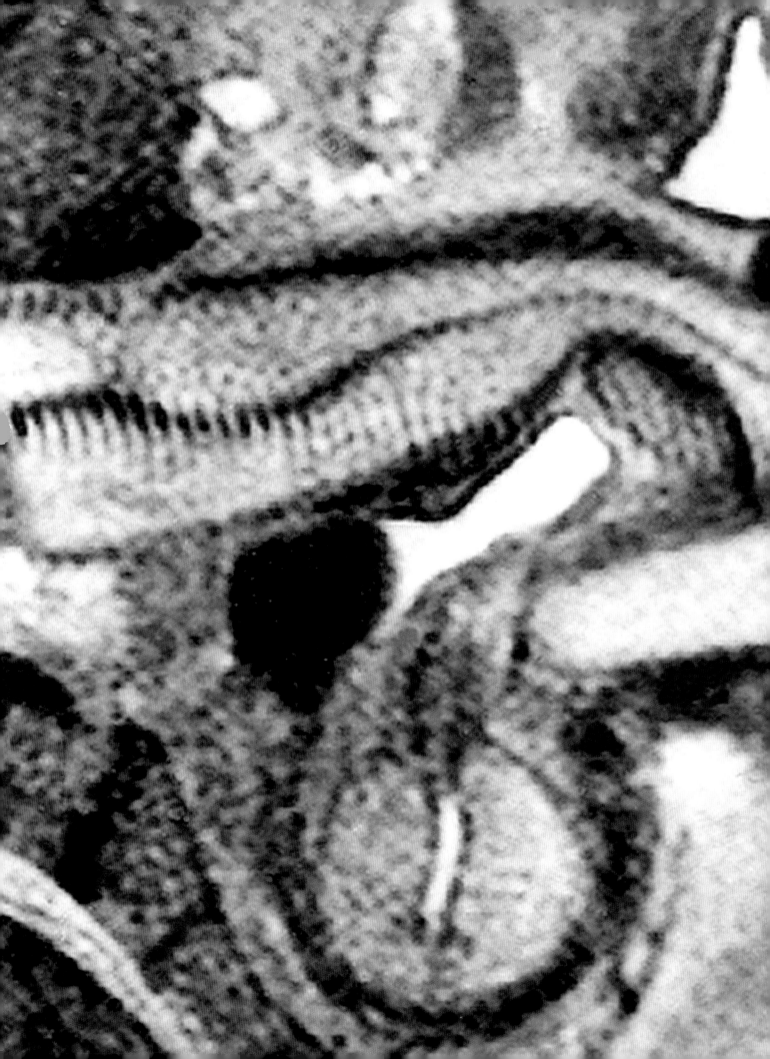

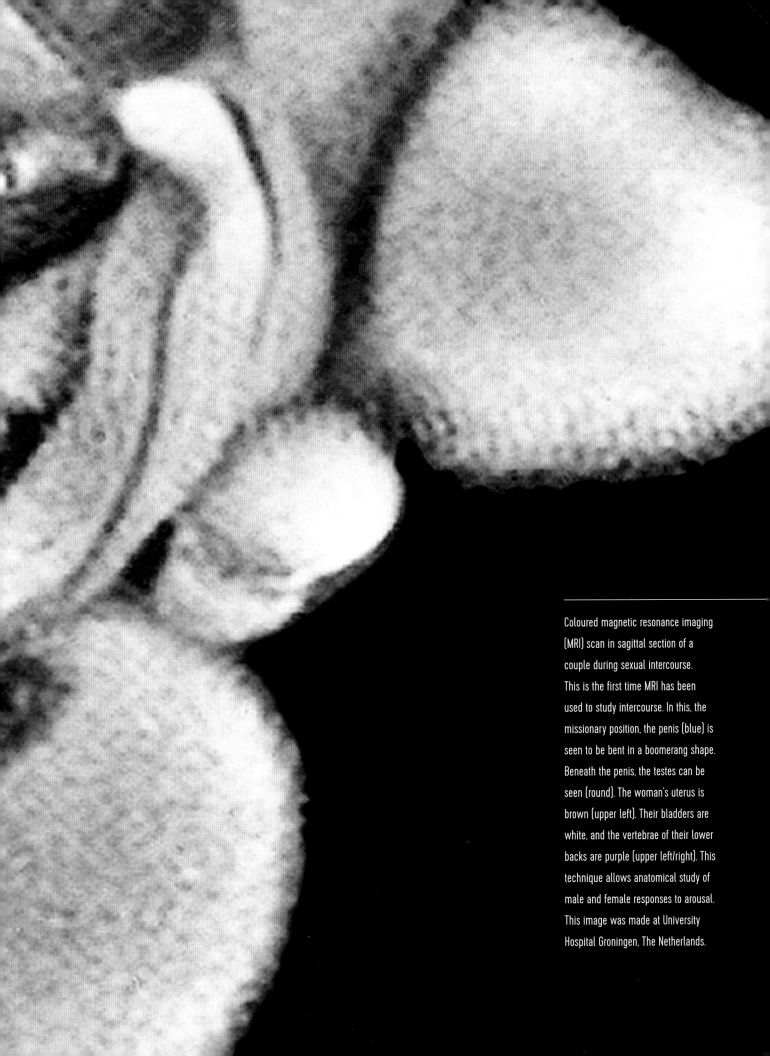

Coloured magnetic resonance imaging (MRI) scan in sagittal section of a couple during sexual intercourse. This is the first time MRI has been used to study intercourse. In this, the missionary position, the penis (blue) is seen to be bent in a boomerang shape. Beneath the penis, the testes can be seen (round). The woman's uterus is brown (upper left). Their bladders are white, and the vertebrae of their lower backs are purple (upper left/right). This technique allows anatomical study of male and female responses to arousal. This image was made at University Hospital Groningen, The Netherlands.

Transmission electron micrograph (TEM) of a section through the nervous system of the intestinal tract, showing a collection of nerve cell bodies called a ganglion. The nervous system of the gut coordinates the functions of the gut including secretions and movement with such complexity it is sometimes known as the second brain.

This inheritance is most clearly demonstrated by the age at which girls get their first periods, a milestone known as menarche, which has been found to have a strong correlation between mother and daughter.

Genetic factors are only part of this story, the environment is playing a major role. One of the most intriguing findings illustrating this takes us back to the childhoods of our great-great-grandparents and even further into the past. Studies looking at the age of menarche in our ancestors have found that intriguingly just two or three generations ago the age of puberty was substantially later – it seems your great-great-grandparents were children for much longer than we are today. In fact in a study of records going back to the mid-nineteenth century, we have found that between 1840 and 1950 there was a drop of four months per decade in the average age of menarche in western European girls. In England that meant that in 1840 the average age of a first menstruation was 16.5 years, but by just after World War II it had dropped to 13.5. The same pattern was repeated across western Europe but intriguingly in Japan the decline happened later and faster, with a drop of 11 months per decade between 1945 and 1975. The upshot of all of this is that the average age for the start of puberty is 10–11 for girls and 11 for boys, and the average age at menarche is 12–13 years, whilst boys ejaculate for the first time on average during their thirteenth year.

Such rapid biological change over the last 150 years suggests an environmental shift that we should see mirrored in the same timescale. As we have already seen earlier in the chapter, with human height variation, we know that during this time frame humans became by and large better nourished, especially through childhood, and it's this nutritional revolution that seems to have driven the increasingly early onset of puberty. Other factors like ethnicity, socio-economic background and exposure to stress all seem to play a part, but overwhelmingly the one crucial limiting factor in the triggering of puberty is the level of body fat. It seems we need to have enough energy in reserve before our bodies are ready to commit resources to the monumental metamorphosis ahead. It's why teenagers involved in elite levels of sport often have a delayed puberty, as do those with anorexia nervosa. It's also why growing levels of obesity in the developed world seems to be driving an even younger start to puberty than we have seen in previous generations.

So it seems the puppy fat of our youth is not just designed to make us feel self-conscious, it is an essential part of our growing-up process, an energy reserve laid down to power us through puberty.

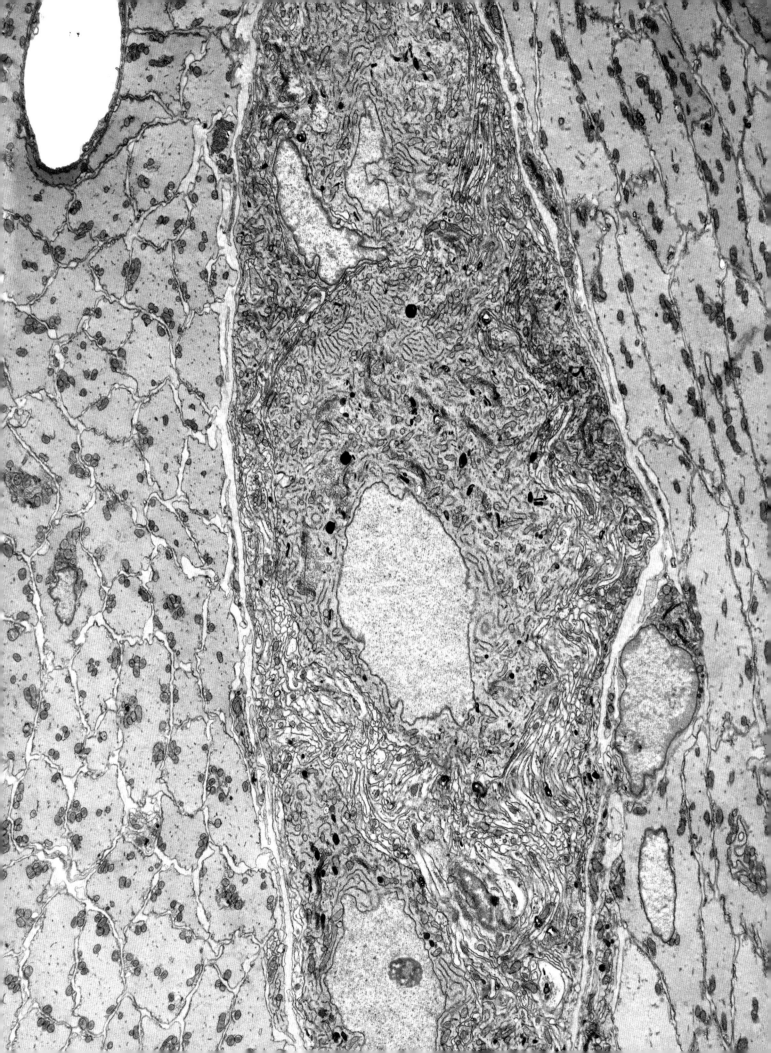

## CLEVER GUTS

Throughout our lives our appetite for food and nourishment feels simple. An unassuming signal we call hunger tells us it's time to eat and then it's up to us to decide whether to pull out the iceberg or plump for the ice cream. It feels as if it's a combination of deep instinct followed by conscious choice. But the complexity of our guts and their relationship with our behaviours is far more influential than we ever imagined. The obesity crisis we see in the developed world has revealed to our collective cost the inbuilt bias we all have for high-energy, calorie-rich foods, but the decisions we make when standing in front of the fridge go even deeper than that.

The human digestive tract is by far the largest sensory organ in the human body: with a surface area of around 32 m² it's lined with sensory receptors constantly monitoring what's going on inside. If you realise that your skin, the second biggest sensory organ, is only 2 m² then you can see just how plugged in the human digestive tract really is.

All of this sensory information is fed to the gut's own brain – the enteric nervous system (ENS) – a network of neurons throughout the entire length of the digestive tract. The ENS uses this information to directly control the muscular action and chemical environment of the gut in order to optimise digestion.

But we've recently discovered that some of the gut's sensors send information elsewhere. At Harvard University in Boston, a team have recently discovered a new class of neurons that directly sense nutrients such as sugars, fats and protein.

Intriguingly these nerves bypass the gut's own nervous system, ignoring the ENS and sending the information directly to the brain. Why the brain needs to know what's going on in the gut so rapidly is unknown but one theory is that it could be used to directly influence our eating decisions, making us eat more when foods are nutrient rich. This would mean our guts are far from just passively digesting what we eat; they might be controlling our behaviour to ensure we get the energy we need and playing a key role in driving our snack-hungry teenagers.

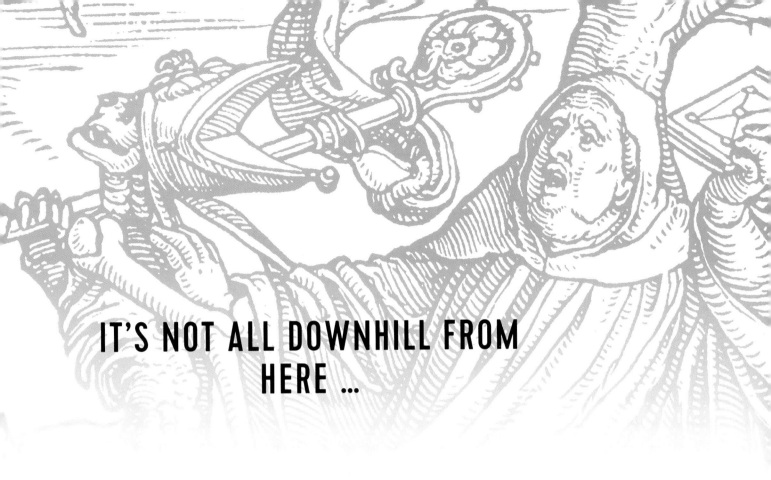

# IT'S NOT ALL DOWNHILL FROM HERE ...

After almost 18 years of constant growth, change and development, the human body is finally fully equipped for its journey through adult life. Emerging from the throes of puberty stronger, fitter and sharper than ever, the twenty something human is in its prime, a body that will in many ways never be better in terms of physical capability. We see this play out most overtly in the careers of elite sports men and women. Footballers reach the peak of their performance in their mid-twenties, elite sprinters around a similar time – in fact as a rule, all explosive sports require a twenty something body to be the very best. Endurance sports on the other hand seem to suit a slightly older body, with peak performance in events such as marathon running, cycling and triathlons all coming later in life, often in the fourth decade. Whether this is the power of the mind driving the body, or simply a physiological peak is still unclear but it's a phenomenon that is seen across a wide range of endurance disciplines.

The news is not so good, however, when it comes to the capabilities of the mind for those of us already past our twenties. The ability to commit new infor-

'I REALLY BELIEVE AGE IS A STATE OF MIND. MOST MEN MY AGE ARE DEAD AND BURIED LONG AGO, IF YOU STAY IN SHAPE AND PUSH YOURSELF PHYSICALLY YOU CAN ACHIEVE AMAZING THINGS.'
**'IRON' LEW HOLLANDER**

mation to our long-term memories is in decline often before we even leave our teenage years, our memory for face recognition peaks around our thirties and it's a painfully often quoted fact that most Nobel Prize discoveries are made before we hit our forties (I have two years and two months to get my act together!). But even with these declining cognitive capabilities, it's not all bad news; experience and wisdom make up for a lot of these deteriorations, allowing many mental skills such as arithmetic, comprehension and social reasoning to keep on improving well into middle age. It's a reassuring certainty that there is no single prime age for the human body, no moment when we are in peak condition for absolutely everything, but even so we do all face a battle to keep our bodies and minds in optimal condition once we tip over into adulthood.

For many people the outward effects of ageing that we gradually see in the mirror and feel in the morning are no longer an inevitability but more like a challenge that needs to be overcome, and nobody encompasses that battle more than 87-year-old ironman triathlete Lew Hollander, or 'Iron Lew' as he likes to be known.

For most people half his age the idea of completing a 2.4 mile swim, followed by a 112-mile bike ride and a full marathon to finish the day off, all in under 17 hours, would be enough to send us running (slowly) for the hills. For Lew, however, it's just another day out: this year has seen him notch up his 100th ironman triathlon, holding multiple records in the over-eighties categories, and that's just one of the many endurance competitions he takes part in.

It's a track record that Lew puts down to a few simple life rules. The first and most important is 'go anaerobic every day', or in other words exercise until you are out of breath. The second is 'use it or lose it' – as Lew wisely declares, 'Don't say, "Ow, my knee hurts". If you get out and use it, your body will tend to make it better again.' And the third and possibly best piece of advice anyone has ever offered is 'Eat dessert first, because life is uncertain.' Although he is also quick to add: 'There are no old, fat people, so watch your calories!'

Lew is as fit an 87-year-old as you are ever likely to meet, but the only reason he is able to compete in such extraordinary races is that most of his body is nowhere near that old. We may appear to stop growing at the end of adolescence, but every one of us relies on an endless cycle of growth and renewal to keep our bodies in working order: our bodies are in a constant state of change. We might not be quite up there with Lew in terms of giving our bodies a real beating but even living a more sofa-friendly lifestyle, the cells that make up our tissues, skeleton and organs are continually wearing out and in need of replacing.

Cell regeneration is a secret part of human growth that happens every day, from birth right up to the moment of death. Almost every part of our body renews itself; as far as we know only a tiny number of tissues such as the lens of the eye are set for our whole lives. We used to think that heart muscle as well as the brain and nervous system were made up of cells that were unable to replace themselves, but even these tissues retain an ability, although somewhat limited, to regenerate throughout our lifetimes.

'Iron' Lew Hollander, powering on into his late eighties.

The rate of tissue renewal varies throughout our bodies (see table). Our skin, the largest and fastest-growing organ, makes up around 15 per cent of our body weight.

| Cell Type | Turnover Time |
| --- | --- |
| Small intestine epithelium | 2–4 days |
| Stomach | 2–9 days |
| Blood neutrophils | 1–5 days |
| White blood cells: eosinophils | 2–5 days |
| Gastrointestinal colon crypt cells | 3–4 days |
| Cervix | 6 days |
| Lungs: alveoli | 8 days |
| Tongue taste buds | 10 days |
| Platelets | 10 days |
| Bone: osteoclasts | 2 weeks |
| Intestine: panath cells | 20 days |
| Skin epidermis cells | 10–30 days |
| Pancreas: beta cells | 20–50 days |
| Blood B cells | 4–7 weeks |
| Trachea | 1–2 months |
| Hematopoietic stem cells | 2 months |
| Sperm (male gametes) | 2 months |
| Bone: osteoblasts | 3 months |
| Red blood cells | 4 months |
| Liver: hepatocyte cells | 0.5–1 year |
| Fat cells | 8 years |
| Cardiomyocytes | 0.5–10% per year |
| Central nervous system | <lifetime |
| Skeleton | 10% per year |
| Lens cells | lifetime |
| Oocytes (female gametes) | lifetime |

Elsewhere throughout our bodies the rate of renewal is fast and furious (see table). Our skin, the fastest-growing organ, comprises 1.6 trillion cells that make up 16 per cent of our body weight. It's our first line of defence from the outside world, so the rate of cell death and regeneration is incredibly high – in fact every month the outer layer of our skin is totally replaced.

Similarly our lungs undergo a barrage of assaults from the outside world as we breathe in 5–8 litres of air every minute. The cells lining the surface of the lungs are the first line of defence against airborne microbes and dust and are replaced every two or three weeks. But deep within the lungs in the alveoli air sacs, where oxygen is extracted from the air and carbon dioxide is expelled from the blood, there is a more steady progress of regeneration that takes about a year. The blood cells that carry the oxygen away from the alveoli and speed around the body's 100,000 km of blood vessels are not so long lived. As these biconcave discs get forced through capillaries as small as 4 microns to deliver their payload of oxygen, it's only their flexible membranes that allow them to squeeze through.

Completing their journey around the body 60 times every hour, the extreme mechanical stress of the voyage ultimately takes its toll and after 120 days they are worn out. The cell membrane that had contorted itself through the smallest of gaps now undergoes a subtle but significant change, marking it out to the macrophages of the immune system. A cell that has served the body so diligently for three long months now becomes a target and an immune response is triggered in the spleen, liver and bone marrow. The old red blood cells are engulfed by macrophages – a process known as phagocytosis.

Phagocytosis breaks down the old red blood cells at a rate of 150 million cells every minute, but deep inside our bone marrow new red blood cells are being produced at an equal rate ensuring the demands of our oxygen-hungry bodies are constantly met.

Even the parts of us that appear permanent are not quite what they seem. Our skeleton, for example, is constantly being broken down and replaced by new bone, taking nearly a decade to completely renew itself. This means that for someone like Lew as he heads towards his ninth decade, he's almost replaced skeleton version eight.

The secret behind many of these processes of renewal lies in the reserves of a special type of cell dotted throughout our bodies, a cell that has been the subject of intense scrutiny for the last few years and which we are only now just beginning to fully understand.

Stem cells were thought of as regenerative cells that existed mainly in our bone marrow, with the specific task of producing new red and white blood cells. We now know they are in fact found throughout our bodies in our skin, gut, liver, brain and bones. As adults we have less of them than we do during childhood, but

THE PROCESS OF PHAGOCYTOSIS BREAKS DOWN
THE OLD RED BLOOD CELLS AT A RATE OF
150 MILLION CELLS EVERY MINUTE

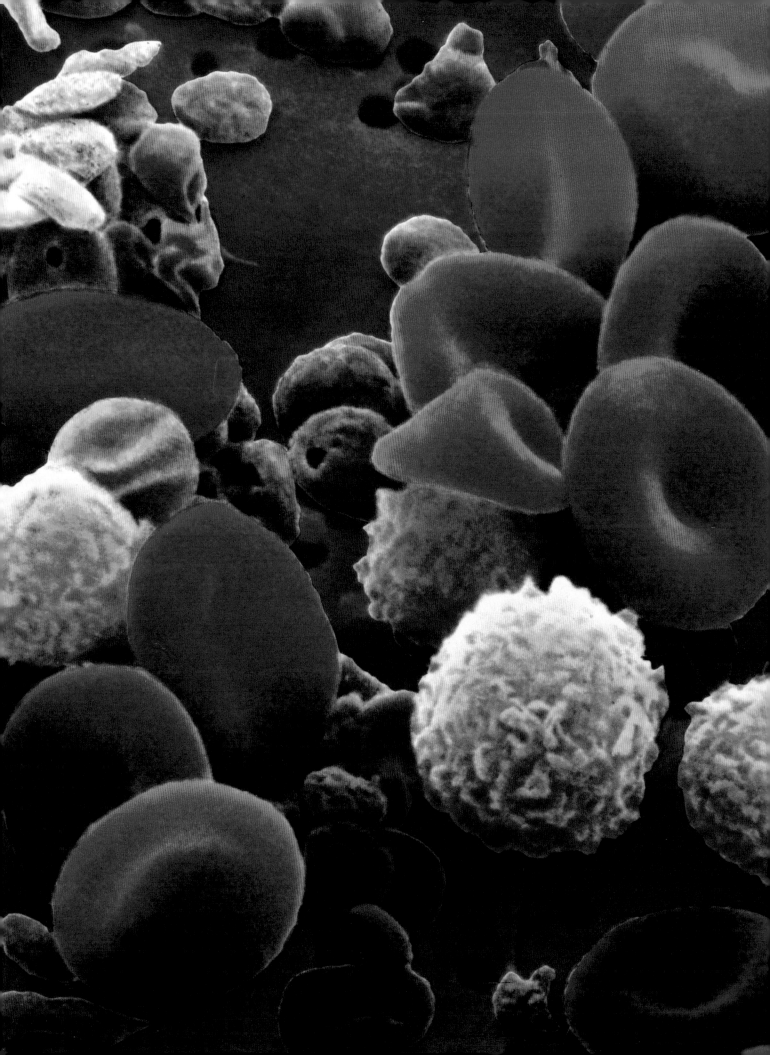

they still manage to grow and generate millions of new cells to replenish and repair our bodies' tissues every single day of our lives. Adult stem cells do not have the property of pluripotence, the ability to turn into any and every cell type. Instead they are limited becoming certain types of cells depending on which type of the three embryonic tissues they arise from.

Unlocking the secrets of stem cells is crucial to the understanding of many diseases and the process of ageing itself. It seems that the impairment of their function as we age is one of the key factors that limits human lifespan to an average of around 80 years. As well as potentially holding the key to an elixir of youth, these cells also offer us other extraordinary possibilities and harnessing their power may give us the ability to control and shape the growth of human tissue in ways that appear more like science fiction. Many believe the stem cell revolution will transform medical care in the twenty-first century. In one lab at Harvard the revolution seems like it is almost here.

**Opposite page:** Coloured scanning electron micrograph of human blood showing red and white cells and platelets. White blood cells (leucocytes) are rounded cells with microvilli projections from the cell surface.

Colour-enhanced image of a capillary. Endothelial cells (purple) line the blood vessel and are surrounded by supporting pericytes (turquoise). Two red blood cells can be seen inside the capillary.

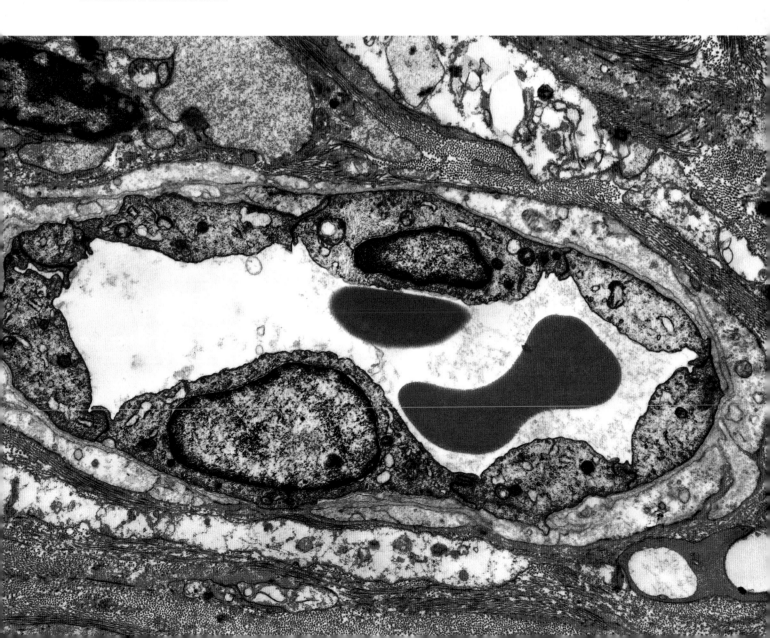

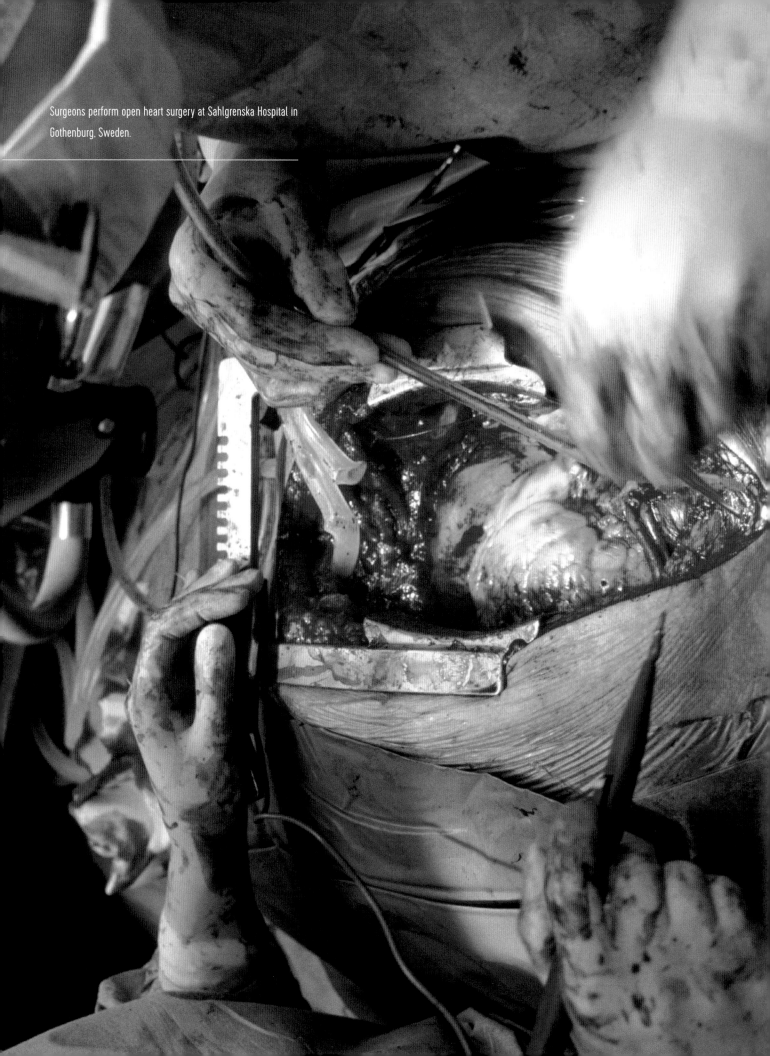

Surgeons perform open heart surgery at Sahlgrenska Hospital in Gothenburg, Sweden.

# ORGANS TO ORDER

When you first walk into Harald Ott's laboratory, it takes a moment to locate what exactly is so odd about the sight of a disembodied heart beating in a glass jar on the bench. It is a modern research laboratory at Massachusetts General Hospital complete with the usual paraphernalia of biological science: pipettes for moving small quantities of liquid around; flasks of nutrient growth medium; plastic and glassware for growing cells and mixing reagents. But there is also a human heart in a large cylindrical glass flask, half immersed in pink liquid, its vessels connected to tubing, twitching or, more correctly, pulsating ever so slightly.

So the immediate oddness is visual: twenty-first-century science with the aesthetic of nineteenth-century science fiction. But there is a deeper oddness too – because this heart is not merely a donated heart kept alive outside the human body. While that would be interesting to see, it would be fairly trivial to accomplish and also fairly pointless. This heart is as close as we have yet come to growing an organ from individual stem cells outside the body. And it's odd, because when I walked into the lab, I was certain it wasn't possible to do that yet.

## THE DESPERATE NEED FOR NEW ORGANS

Growing organs outside the human body is a scientific holy grail. It's hard to think of many advances that would revolutionise medicine as much as organ regeneration. And this is particularly true when it comes to replacing the heart. Other organs can be replaced with transplants from living donors (like bone marrow), machines (haemodialysis for kidney failure) or even drugs (insulin for a failing pancreas). But the heart is needed every second of every day. And it requires another person to die in order to replace it, and in a way that hasn't damaged their heart. A heart from someone who has died of a heart attack is of little use.

There is a dire shortage of hearts and other organs for transplant and a growing demand. In the UK around 3,000 transplants of all organs are performed per year but around three people die per day because of a shortage of organs. In the USA the numbers are even starker. Twenty-two people die each day waiting for a transplant. Around 90 per cent of people support donation but just 30 per cent

**WHAT IS HEART FAILURE?**

As so often in medicine we only appreciate the complexity of a system when it begins to fail. In the case of heart failure, as the pump stops working, organs start to receive less blood and the body tries to compensate. The kidneys respond as they would to blood loss, and start to retain salt and water. This essentially increases the amount of blood in the body and increases the amount of blood flowing into the heart. The heart, being elastic, responds rather like a balloon. If a balloon is inflated a little and released it will simply puff the air out. Blow it up to near bursting point and the energy stored in the stretched walls will blast the air out forcefully. So an 'overinflated' heart will expel more blood, giving the body some temporary respite. But you can't keep filling up the system for ever. The increased water and salt in the blood start to leak into the peripheral tissue and lungs. The patient will report swollen ankles and breathlessness as they lie flat. And blood tests will start to reveal that other organs are being chronically underperfused.

A photograph from John Gurdon's Nobel Prize lecture. A clone of albino male frogs obtained by transplanting nuclei from cells of an albino embryo to enucleated eggs of the wild-type female shown. The albino frogs are genetically identical and will accept skin grafts from each other. This was the first proof that all the information required to grow an adult was contained in a nucleus and that genes were not irreversibly switched off.

are on the register, so that of the roughly half a million annual deaths in the UK only about 3,500 become donors. This puts us all in a situation that should be of great personal interest to each and every one of us: mathematically you're far more likely to need a transplant than you are to become a donor. In the UK the number of heart transplants has halved in the last 10 years for unknown reasons. It may be because we're squeamish about persuading families of brain-dead patients to donate. It may be because road safety has increased, but whatever the reason there is now a desperate shortage. This is worsened by a rise in the number of people who need heart transplants.

This increasing demand is paradoxical, because doctors have become better at treating heart attacks. The most common reason a heart transplant is needed is because of a condition called congestive heart failure. Heart failure simply means that the pump of the heart has started to fail; it can no longer supply the blood requirements of the body. There are lots of causes but the most common is a heart attack or myocardial infarct, where a clot forms in one of the arteries supplying the heart muscle itself. Cardiologists have had immense success in treating heart attacks in the short term. There are drugs which break down the clot and it is now standard treatment to guide an expandable balloon into the blocked artery, usually via a wire inserted into the groin. These interventions mean that the death rate has plummeted in the last few decades. But the result is that more and more people survive with damaged hearts that gradually start to fail. The cardiologists have in a sense become victims of their own success. Heart failure is very serious. Some 30–40 per cent of patients diagnosed with heart failure die within a year and 10 per cent more continue to die each year after that. This makes it a more dangerous condition than a typical cancer.

With increasing demand and decreasing supply, the limited number of organs puts an ethical strain on the transplantation teams. Only the sickest patients get new hearts, those who cannot wait any longer. But they are the ones least likely to survive what is a huge operation. While the old heart is removed and the new one sewn into place, the whole body must be temporarily plumbed into a machine which acts as the heart and the lungs. By the time an organ is identified, for many people they are too sick for the operation.

And even if you get a transplant, life is far from simple. Your body is exquisitely sensitive to non-self – this is the foundation of immunity: the ability to detect something alive that is not you … and kill it. Even if the donor is a close relative, the recipient immune system recognises the transplanted tissue as 'non-self' and quickly destroys the transplant. The only way of stopping this is with powerful, life-long drugs that suppress the immune system. This of course comes with a range of serious side effects, including increases in infections and cancers.

So the goal of the Ott lab is to be able to grow an unlimited supply of new, young hearts (and other organs) that are perfectly tissue-matched to a patient.

## HOW TO GROW A NEW ANIMAL FROM A SINGLE CELL

The story of the heart in front of me on the lab bench starts in 1956, when a graduate student at Oxford named John Gurdon started trying to swap nuclei between frog cells.

The history of biological science is full of geniuses who have spent their life working on projects that seem unimaginably arcane. To understand why Gurdon did these experiments, and at the risk of me sounding like a parent having 'the conversation', here's a little biology revision. Your entire body came from a single

**BROTHER CAN YOU SPARE A LUNG?**

If like me you have an identical twin with whom you share 100 per cent of your DNA, then a tissue match will be perfect but even having a twin isn't foolproof. If I caught him in a good mood my brother might be prepared to give up a kidney or some bone marrow. But he's unlikely to be talked into donating a full set of heart and lungs.

cell called a zygote, which was itself a fusion of two cells: a sperm and an egg ('Mummy and Daddy had a very special feeling ... etc'). Now, the sperm and the egg each have half the amount of DNA of a normal human cell, 23 chromosomes, so the zygote ends up with the full complement of 46 chromosomes, half from each parent. Every cell in your body comes from this single zygote cell, which we describe as totipotential: it has the potential to become all the different cells in your body.

Every time the zygote, and all its trillions of daughter cells divide, they replicate all their DNA. So every cell in your body contains all of your genes. Understanding this is at the centre of Harald Ott's research and was what John Gurdon was trying to prove in 1956 with frogs. It was known that different body tissues in either frogs or humans are specialised. They make different proteins and this process is delicately controlled. The gut and pancreas produce digestive enzymes, for example. But if these same enzymes were produced in the brain, they would dissolve it. This is an extreme example, but specialisation of cells is what enables multicellular organisms to function, and we now know it comes from genes being switched on and off.

And specialisation starts early. After a few divisions the zygote divides into a ball of around 300 cells called a blastocyst and it is this ball that implants in the uterus. Even at this early stage the fates of cells start to be determined. Some cells around the outside of the ball go to make up the tissues of the placenta (the trophoblast), while others become the inner cell mass. This cell mass is made up of embryonic stem cells. These cells can become any cell in an embryo ... but they're already specialised and committed – they can't become placenta. Or rather they won't. They contain all the same DNA as the cells of the trophoblast that will become placenta but some genes are, in nature at least, irreversibly switched off.

But at the time Gurdon started work with frogs, it was believed that cells might be discarding genes, getting rid of them permanently on their journey to specialisation. Gurdon's supervisor suggested that he try to replace the nucleus in a frog zygote with the nucleus from a mature cell from another frog's gut. He chose the South African clawed frog *Xenopus laevis* as it has an astoundingly large egg, clearly visible to the naked eye with a 1.2 mm diameter, about ten times the width of a human egg (or several hundred times the width of a red blood cell).

Gurdon expected to show that a nucleus from a mature frog cell was indeed missing genes. It took him a year to perfect the technique required to penetrate the thick slime surrounding the egg with a tiny pipette – in his Nobel lecture, Gurdon explains in some detail the technique required to penetrate the slime. After hundreds of experiments he was able to show that the nucleus from a mature frog gut cell could be sucked out and put into an egg. This would then develop into an entirely new frog. It was the first example of vertebrate cloning in a laboratory: making an entirely new animal genetically identical to another. It would take 40 years before the same experiment was repeated in a mammal, Dolly the sheep.

So Gurdon proved that cells can become entirely specialised without losing any genes permanently. The genes are reversibly switched off. From this moment the possibility of growing organs in jars on lab benches moved from fantasy to dim possibility. In principle it would be possible to take a cell from an adult human and turn back the clock on specialisation and start again. This opened up the field of stem cell research.

## TURNING BACK THE CLOCK: MAKING STEM CELLS FROM SPECIALIST CELLS

From the first specialisation step where cells are destined to become either placenta or embryo, cells become increasingly committed to their tissue types. The cells in your body can be roughly divided into stem cells and cells which are specialised or 'terminally differentiated'. They have a single job and will not divide any more. Some are extremely long-lived like neurons and eggs in women, which last a lifetime. Others like sperm, skin and gut cells have a lifetime of a few days before they are replaced, and they are replaced by stem cells dividing. There are lots of types of stem cells, each found in different tissues. We hear a lot about embryonic stem cells in the press, perhaps because of the ethical dilemma they present for some people: they need to be harvested from foetuses that have been terminated. But of greater interest to many stem cell scientists are the specialised stem cells found in each tissue. There are stem cells in your bone marrow which divide to make blood, stem cells in your skin which constantly replace the layers that slough off and stem cells in your muscle called satellite cells which give rise to muscle cells.

Gurdon shared his Nobel Prize with Shinya Yamanaka, a Japanese scientist born the year of Gurdon's main discovery. Forty years after the first successful frog nucleus transplant, Yamanaka astounded the scientific community when he showed that cells in the skin could be turned back into pluripotent stem cells, able to produce any tissue of the body, by manipulating four genes. These are called induced pluripotent stem cells and they bring us back to the Ott lab.

*DO ALL CELLS HAVE THE SAME DNA IN THEIR NUCLEUS?*

There are a few exceptions. Red blood cells don't have a nucleus. White blood cells reorganise their DNA in order to produce the massive variety of genes necessary for constructing antibodies and the receptors they use to recognise different microorganisms.

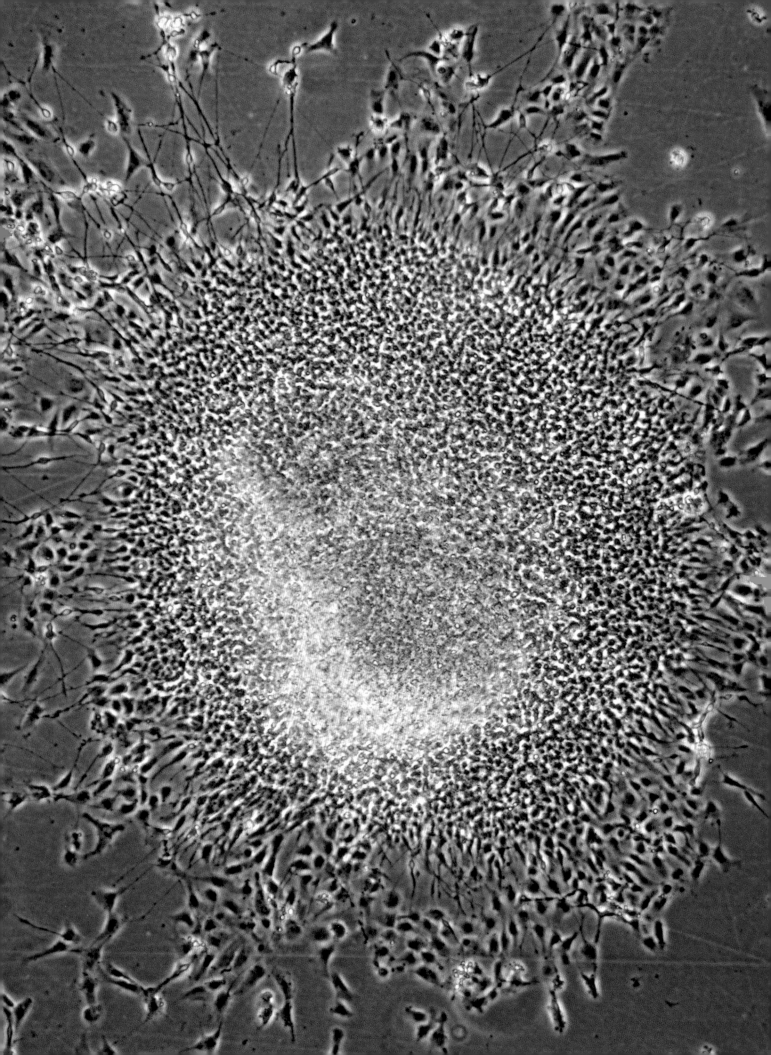

# TURNING THE CLOCK FORWARD: FROM STEM CELLS TO SPECIALIST CELLS

The story of the work done in the Ott lab is the story of the secret complexity of the human body, complexity that is only revealed when we try to replicate what the body does. Why, over 60 years after the discovery that mature cells could be induced to regrow whole organisms, do we not have an unlimited supply of tissue-matched hearts? Why can't we just take a cell from an adult, reprogram it to become a stem cell and grow a new heart? This is the question the Ott lab continually butts up against as they inch towards their goal.

Firstly, just because you can create a stem cell from, say, a skin cell, does not mean it's easy to then turn it into a heart muscle cell, although Harald is blasé about this. 'You just put a few chemicals in the dish with the cells and they become cardiac myocytes, it is not hard.' To be clear, it is hard, it's just a lot less hard than anything else he does in the lab. Harald is also one of those people for whom the word 'hard' seems to be missing from their vocabulary. Things are described as challenging but never hard. It's worth mentioning that throughout our chat Harald was wearing surgical scrubs and carrying a bleeper. As well as running a lab at the cutting edge of modern science, he also operates on patients.

But while growing a dish of twitching cardiomyocytes (heart muscle cells) may look amazing (check out the videos on the internet if you don't believe me), they're a long way from an organ that can pump around 10 litres of blood around your body every minute for up to a hundred years.

The question was how to get a single layer of cells in the dish to take the shape of a heart. And more than that, how to get the cells to have the detailed microscopic architecture that they needed to function. Scientists had come up against this problem previously when they had tried injecting stem cells into damaged hearts after heart attacks – the stem cells would stick to the dead heart but since new blood vessels didn't form and run through the new tissue, they would die.

This problem of how to get the heart cells to take the form of a heart gets to the core of what you are actually made of and how DNA encodes for you. I often teach science in schools, and teens are almost universally familiar with the idea of cells – the story of van Leeuwenhoek examining his own sperm in the late seventeenth century under the first microscope sticks in the collective imagination – but how does a collection of gelatinous cells become a mobile, robust human? Or a tree for that matter? The answer is the extracellular matrix.

Chinese scientists have published a method to produce multipotent stem cells from human urine. Epithelial cells from the kidneys that pass out of the body with urine are used to produce the stem cells. These cells are capable of developing into any other cell type, and could potentially develop into whole organs.

# THE BODY'S LEAST CELEBRATED COMPONENT: THE EXTRACELLULAR MATRIX

Cells build their own protein scaffold to provide structure to the body. This is the extracellular matrix, perhaps the least celebrated of all your body's components. It barely has a category. It's not quite tissue but it is in all tissues. It's not an organ itself but it gives shape and structure to your organs, tissues and bones. The extracellular matrix is secreted by your cells and comprises a diverse mesh of long protein and sugar molecules which form a framework throughout the body. If we dried you out completely, about 75 per cent of the dry weight of your body would be extracellular matrix protein. In some tissues there is more of it than in others. Around 80 per cent of your skin and tendons is made of extracellular matrix. Bones and teeth are made of matrix which has become calcified, and your cornea, the clear tissue at the front of your eye, is made from transparent matrix. Basically, the extracellular matrix is what stops you existing as a single layer of slime on the earth's surface.

The matrix has two main components. Firstly there are branched chains of sugar molecules bound to proteins called proteoglycans, which form a gel. This gel resists compression and allows nutrients, and waste products, to flow through it to the blood vessels. Secondly, there are fibrous proteins that give strength and allow cells to stick down. For example, collagen fibres provide enormous strength and elastin fibres allow stretch. Just as in stretch jeans the strong fibres comprise about 95 per cent of the tissue, whilst a tiny proportion of elastic lycra fibres allow resilience.

The types of proteins and sugars vary from place to place giving each tissue its particular structural properties or textures. If you eat meat, you will be intimately familiar with the extracellular matrix of vertebrates. Chewy tendons are comprised mainly of the aligned fibres of type 1 collagen. Gristle is the cartilage, made of a different type of collagen, more rigid, resistant to compression. It is what gives the exquisite crunch to a pig's ear, if you're into that kind of thing. The perfect ratio between meat and the extracellular matrix is found in a thick rib-eye steak, cooked medium rare.

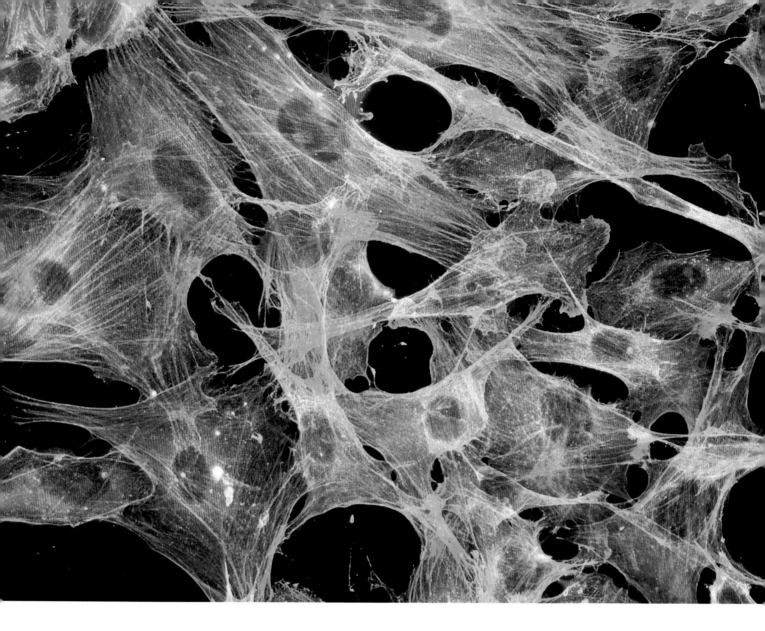

Pescatarians will be more familiar with the calcified extracellular matrices of crustaceans and molluscs as they battle their chitin exoskeletons and shells to extract the more cellular muscles of these organisms.

Cellulose made of long chains of sugar molecules, meanwhile, is the plant equivalent of collagen, giving structure and rigidity to higher plants. If you consider an alga compared to a sequoia, you get an idea of what we would look like without an extracellular matrix, and it's not pretty.

The heart has a highly organised extracellular matrix that the muscle cells, the cardiomyocytes, live on. It forms a kind of organised internal tendon. Tendons normally connect muscle to bone, but the heart is unusual as a muscle in that it doesn't move any bones, just blood. Harald Ott compares it to a building. 'If we consider the people who work in a building as cells, then the building itself is the extracellular matrix.' It was previously thought that the matrix provided merely an inert physical structure, but it's now known that especially in the heart it is

Immunofluorescence light micrograph of human osteosarcoma cells stained to show cell proteins and organelles. The stains used here have highlighted the intracellular structural protein actin (green). Extracellular matrix protein filaments (red) are secreted during wound healing and tissue repair. Osteosarcoma is a type of bone cancer.

dynamic and constantly changing. As well as being a physical structure, just like a building it has a massive range of sophisticated functions: plumbing, electrics, communication cables, waste disposal. It allows the concentration of molecules needed for signalling between cells, activation of enzymes, it provides guidance for cells about where they should sit, and allows the diffusion of waste and nutrients through it. It also concentrates the forces generated by contracting muscle cells in the precise directions needed for the most sophisticated pump in the universe.

Understanding how the extracellular matrix is created is not trivial. It is a process somewhat equivalent to picking oneself up by one's own boot straps. It is secreted by the cells as they develop in the growing embryo but then provides the structure for more cells to grow on. Collagen is the major fibrous protein in the heart. It is secreted by specialised cells called fibroblasts. But the fibroblasts don't just ooze collagen, they then crawl over it dragging and pulling it into sheets and fibres. The fibroblasts organise the collagen and the collagen organises the fibroblasts.

## THE HEART AS A PUMP

The heart's sophistication as a pump is really down to the matrix. Muscle cells are remarkable – perhaps because you can see them actually move under a microscope – but it's the matrix doing the work. Remarkably, after 500 years of dissecting hearts, there is a massive amount we still don't know about how it works. Some of the most useful anatomy has involved pulling apart hearts by hand – when this was done it revealed that the heart is not a muscular bag so much as a wrap. It doesn't so much squeeze blood out as wring it out like you might water out of a towel. This explains how it is able to perform a minor miracle. Heart muscle fibres are only able to shorten their length by about 20 per cent. Yet the heart is able to reduce its interior volume by 60 per cent. By wrapping heart muscle cells around so they pull diagonally against each other along sheets of connective tissue, they twist the heart empty. The elastic matrix then rebounds, sucking more blood into the heart. This 'sucking' is controversial but there is good evidence for this from the same scientist who named the coil of the heart 'Buckberg's coil' – one Gerald D. Buckberg. The heart muscle does not get its oxygen from the blood flowing through it, it has the same fine network of tiny blood vessels or capillaries supplying it as all other tissues of the body. It is fed by two arteries (the coronary arteries) at the base of the main blood vessel leaving the heart (the aorta).

**Opposite page:** A partially recellularised human whole-heart cardiac scaffold, reseeded with human cardiomyocytes derived from induced pluripotent stem cells, being cultured in a bioreactor that delivers a nutrient solution and replicates some of the environmental conditions around a living heart.

So while the embryo is growing, the cells of the heart lay down the matrix and the matrix forms a scaffold for the heart cells. This growing structure is further acted on by the forces generated by the heart beating and pumping blood. This starts to happen just 22 days after conception while the embryo is about the size of a lentil.

And this was the crucial barrier to Ott's team. How to shape a heart from the beginning and direct the precise organisation of cells and matrix. Instead of trying to direct this, Ott pioneered a technique to get around the problem, now affectionately known as decell recell, short for decellularisation, recellularisation.

Developed in 2008 it involves perfusing the arteries of a donated heart with a weak mixture of detergents, the kind you find in shampoo. The walls of cells are made of fats. Detergents dissolve fats in water, leaving proteins behind. Starting with rat hearts, the team have now perfected the technique. A decellularised heart is a flopping ice-white facsimile of a heart, intact in every detail but without any cells. This sounds simple but it is miraculous to me that it could possibly work. The microscopic blood vessels, the delicate signalling proteins, the organisation that allows the heart to beat and pump – all of it is left intact and ready to be repopulated with new heart cells grown from stem cells. Astoundingly the team have done it with other organs with far more fragile structures. Sarah Gilpin, one of the senior scientists in the Ott lab, has been decellularising lungs. The extracellular matrix membrane that cells in the lungs sit on is fantastically thin in order to allow gases to pass across it quickly. It seems impossible to believe it would survive being continuously washed with detergent to remove the cells, but her research has reached a point where they are able to put a recellularised lung back into an animal and have it survive.

So the heart beating on the bench has a donated protein skeleton, stripped of all cells. This framework has been painstakingly injected with heart cells grown from stem cells by one of the post-doctoral scientists at the Ott Laboratory for Organ Engineering and Regeneration. The key thing here is that the matrix is not recognised by the immune system. So a heart from an unmatched donor, or even a pig, could be stripped of cells and then repopulated with stem cells taken from the patient's body and turned into heart cells.

The difficulty now is that under the microscope these cells look like the heart cells of an infant. They're poorly coordinated and extremely weak. Whereas an average heart can generate a blood pressure of 130 mm of mercury – enough pressure to squirt blood from heart height to a low ceiling – this heart can only generate 1 mm of mercury pressure. It can squirt blood just over a centimetre. It is essentially an entirely untrained heart. Harald hopes that by growing the heart

with a combination of the electrical signals that a heart would receive and subjecting it to the physical forces of flowing blood, he will induce the immature heart cells to transform into a heart that beats powerfully and can be transplanted.

It is astounding work, encompassing a broader understanding of the structure of a human being than anything I've ever observed. From the processes that shape and specialise our cells, through to the principles of self-organisation that govern the growth of a new human from a single cell. The goal of a transplantable heart may not be achievable for decades but along the way work like this is helping us to understand the detail of how invisible molecular processes make very visible living human beings.

Harald Ott with a twitching decellularised heart on his lab bench. A donor heart has been stripped of cells with detergent and then populated with stem cells.

# THE ELIXIR OF YOUTH

As we grow upwards and outwards we also grow forwards through time: we grow old. As we age, other forms of growth slow and gradually cease entirely. It's not clear why it is biologically advantageous to start again every few decades, reproducing yourself by diluting your genes 50:50 with those of someone else, rather than say living for centuries, like some reptiles or sharks, or millennia like some plants do.

Efforts to delay or even reverse ageing have centred around our understanding of the mechanisms that make us grow in the first place. The same processes that make us grow also make us grow old.

This seems particularly true for growth hormone. In many respects it seems like a wonder drug. There is a vogue among Silicon Valley tech billionaires to stay young with growth hormone supplementation. In athletes, it is used to increase performance and is a banned substance in all sports, and there are clinics across the United States prescribing it to men in late middle age in desperate hope of regaining their youth. But growth is a two-edged sword. Growing requires cell division, something your body regulates extremely tightly. Uncontrolled cell division is cancer. Dwarfism, where growth hormone is lacking, or ineffective, is associated with reduced cancer risk. Meanwhile in acromegaly, where there is an increase in growth hormone levels, there is also an increase in cancer risk.

Decent studies are lacking but it is widely suspected that supplementing the body's growth hormone levels to fight the natural decline as we age may increase the chance of death from cancer. As usual the body has evolved an exquisitely balanced system, trading off risks of uncontrolled cell growth (cancer) with the need for size and strength.

Once again, in our efforts to improve on nature, we are left in wonder at the solutions it has found to the extraordinary challenge of growing a working human body from a single cell.

Xand, possibly about to do some decorating?

TWO
—

# LEARN

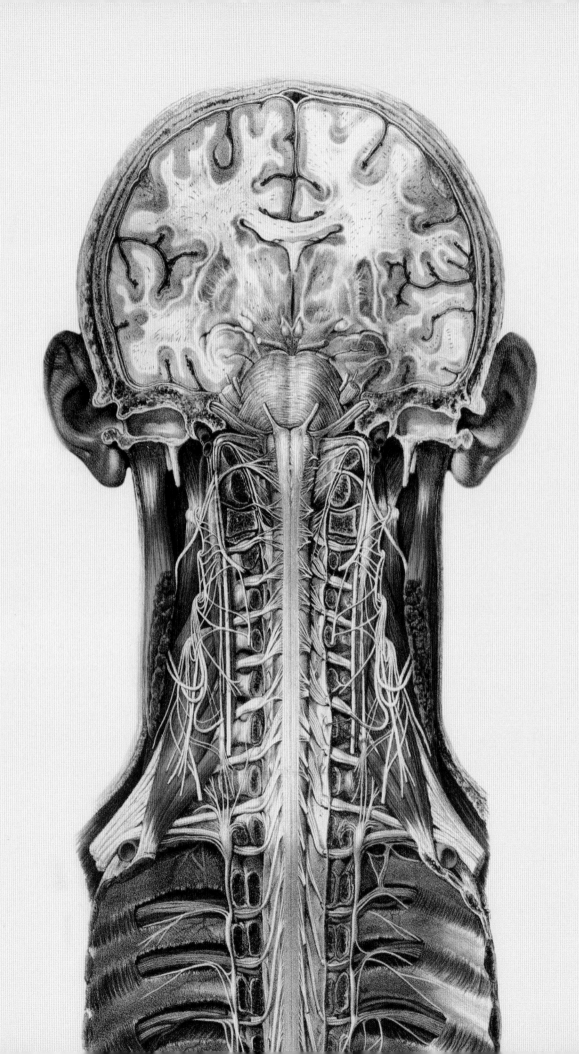

# A LIFE WELL LEARNT

Humans are pretty inept at birth. Unable to walk, talk or feed ourselves, we are born at the mercy of those around us. It's an oddly human thing to be quite so useless. Many animals can walk pretty much from birth. Horses are up on all fours within the first hour and yet for us it is often more than a year before we take our first steps. Even amongst our closest relatives we are born at the back of the class. Primates like chimpanzees and bonobos are born far more agile and able, with brains that are significantly more mature than our own, compared to the adult of the species. In fact it's been suggested that the human gestation period would have to at least double to 18 months for us to be born at a comparative level of maturity. Why we are born quite so prematurely is still a matter of much debate and conjecture.

A long-standing theory, sometimes referred to as the *obstetrical dilemma*, suggests that the 'early' birth of human babies is the result of an evolutionary trade-off between the shape of the human pelvis and the size of the baby's brain. The theory states that to be able to walk and run effectively on two legs (a uniquely human trait amongst our primate brethren), the human pelvis needs to be as narrow as possible. But this narrower pelvis requires a small head at birth. In this evolutionary tug-of-war between mobility and maturity, a balance is struck

**Previous spread:** White-matter fibre architecture from the Human Connectome Project Scanner dataset. The fibres are colour-coded by direction: red, left–right; green, anterior–posterior; blue, ascending–descending.

**Opposite page:** Central nervous system, historical anatomical artwork, from the nineteenth-century French textbook *The Atlas of Human Anatomy and Surgery* by J. M. Bourgery and N. H. Jacob.

Lyra throws her hands out in the classic Moro reflex, one of a range of primitive reflexes that babies are born with. Gradually development of the frontal lobes suppresses these reflexes, but they may return in adulthood or old age with frontal lobe damage or dementia. The Moro reflex occurs when the infant is startled. The arms are flung out and then drawn closely back to the body. It is hypothesised to help infants cling to their mothers but I have been unable to see how, from observing my own daughter, it would do this!

where the female pelvis has narrowed. As a consequence our offspring need to be born earlier and more inept, with a brain size just 30 per cent of the adult brain compared to, for example, the 40 per cent that a chimpanzee infant is born with. In different forms this theory has endured since Sherwood Washburn, an American anthropologist, first proposed it in the 1960s.

Today, however, there is a competing theory that is beginning to gain considerable traction. Studies conducted by Professor Holly Dunworth and her team at the University of Rhode Island in the United States suggest that the length of gestation in humans is defined by the energetics of pregnancy, not the anatomy. If Dunworth is right, then birth is triggered around the nine-month mark because this is the point at which the mother's metabolism can no longer support the demands of the baby. 'What we found is that babies are born when they're born because mother cannot put any more energy into gestation and foetal growth,' she explains. 'Mom's energy is the primary evolutionary constraint, not her hips.'

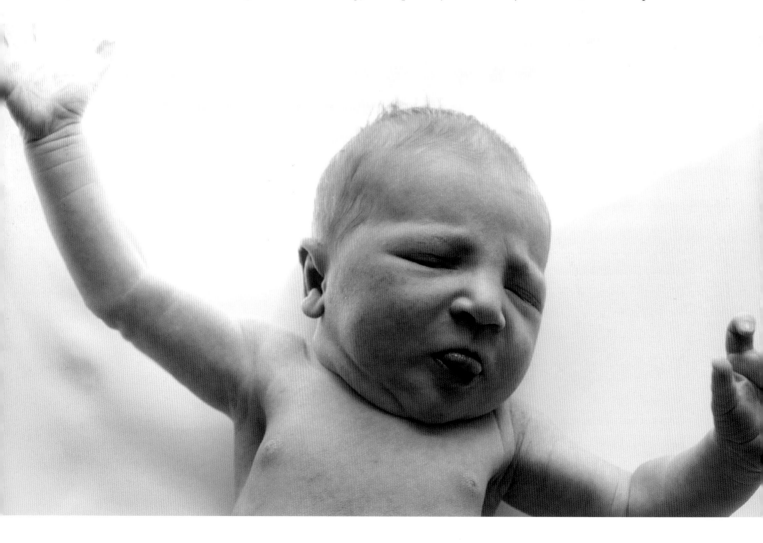

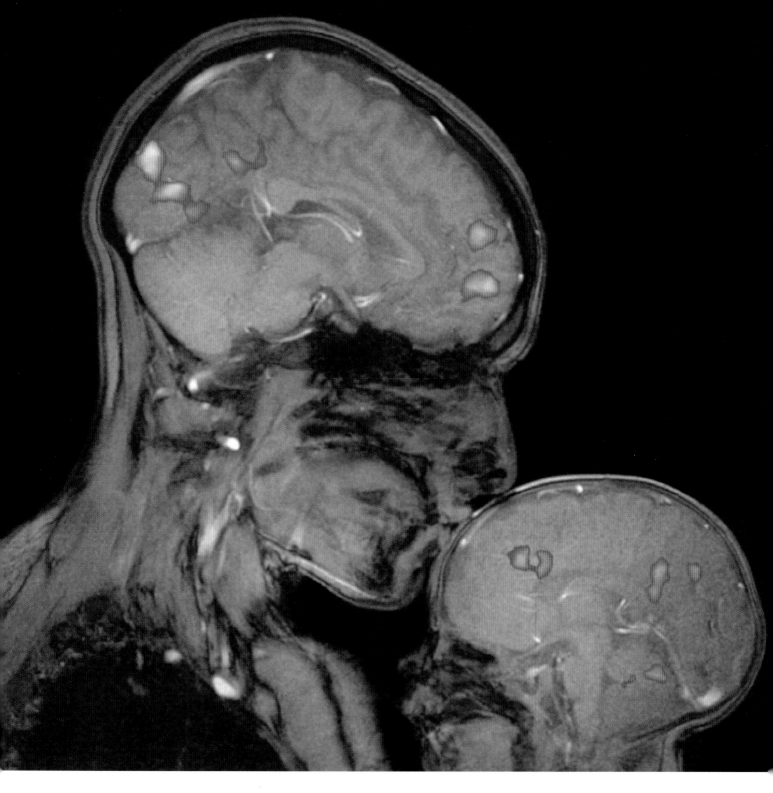

It's an interesting new theory, but as yet there is still no consensus explanation for the obstetrical dilemma. What isn't in doubt, however, is the consequence of the 'early' birth of humans. We, more than any other animal, are born a blank slate, a *tabula rasa* upon which the most extraordinary journey of learning can take place. At birth our abilities may be limited, but the brain we are born with is geared to rapidly suck in the world and transform us into the most intellectually and physically agile creature on earth.

MRI scan of neuroscientist Rebecca Saxe cradling and tenderly kissing her 2-month-old son Percy in 2015. It is possible to see that the infant brain almost entirely lacks myelination.

# FIRST STEPS – EASY RIDER

Riding a bike is one of those skills that we all recognise as having a distinct before and after. Before, it seems like an impossible, complex task outside the capability of our body and brain, but afterwards it becomes second nature, an action that is automated and effortless, an unconscious skill. None of us think about the complexity of riding a bike once we have mastered the task, in the same way as none of us give a second thought to putting one foot in front of the other when we walk. We don't work at the skill, we just do it. But even simple skills like riding a bike are incredibly complex, multifaceted activities requiring a huge amount of coordination. The question is how does something so difficult become so easy? The answer is, of course, practice but the biological process that underpins the journey from clumsy novice to accomplished pro is only just giving up its secrets.

Danny MacAskill is possibly the most skilled cyclist on earth. A love affair with two wheels has taken him on a journey from Edinburgh bicycle mechanic to internet icon. His first film, entitled *Inspired Cycles,* appeared in 2009, 5 minutes and 37 seconds of roughly shot man and bike filmed on the streets of Edinburgh; it's now been watched (at the time of writing) more than 37 million times. Danny makes it look effortless, performing jaw-dropping stunts as if the bike has become an extension of his own body.

Danny MacAskill returned to his native home of the Isle of Skye in Scotland to take on a death-defying ride along the notorious Cuillin Ridgeline.

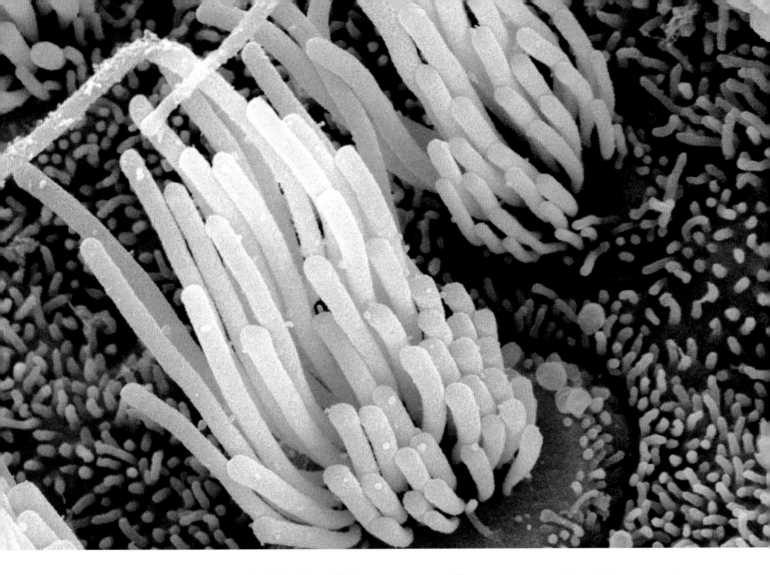

To watch Danny navigate his bike through the most extreme of environments is impressive enough, but deconstruct what is really going on inside his body and the depth of his skill becomes even more apparent. From head to toe multiple systems are integrating together to allow Danny to coordinate his body and bike through space and time. As Danny's hands grip the handle bars and his feet hit the pedals, 50 touch receptors per square mm of skin send signals to his brain about the shape, size, and texture of what they're touching and whether it's moving. His balance system is also constantly monitoring where his body is in space through a process of position detection, feedback and adjustment using communication between the inner ear, eyes, muscles, joints and the brain. While one part of the inner ear enables hearing, another part is designed to track information about the position of his head. The vestibular system in each inner ear is made up of three semi-circular canals and two pockets, called the otolith organs, which together provide constant feedback to the cerebellum about head movement. Movement of fluid inside the canals caused by the motion of his head stimulates tiny hairs that send messages via the vestibular nerve to the brain's movement control centre, the cerebellum. Using this feedback from the environment, Danny's brain sends

The vestibular system in each inner ear is made up of three semi-circular canals: movement of fluid inside the canals stimulates tiny hairs that send messages via the vestibular nerve to the brain's movement control centre, the cerebellum.

■

'WHEN I'M RIDING I FEEL AT ONE WITH THE BIKE –
DON'T EVEN HAVE TO THINK ABOUT IT ...
I CAN JUDGE DISTANCES, HEIGHTS AND BALANCE
US BOTH REALLY EASILY ... I KNOW WHERE I AM
IN SPACE – WHERE THE BIKE IS – AND WHERE
WE'LL BOTH END UP.'
**DANNY MACASKILL**

■

messages to his muscles to move and make the adjustments to body position that will maintain his balance and coordination.

Although, in Danny's case, all of this coordination allows him to control his bike in the most extreme of circumstances, it requires a similar level of complexity just for you and me to ride a bike down the road. And all of that complexity is hidden by a process that makes it feel simple, and it happens in each one of our brains, whether it's learning to ride a bike or walking.

Learning to ride a bike is a remarkably strange process to observe. One minute you are all wobbles and crashes, the next minute you're cruising and in control. This process of learning takes you from an intense conscious effort during the learning phase to an effortless, unconscious activity once the learning is complete. It really is like flicking a switch, but what is that switch in biological terms?

The answer it seems is to be found in a part of your brain that used to be considered somewhat unimportant. The white matter of your brain was long assumed to be of secondary importance to the grey matter – the supposed real substance of thought. Lodged away underneath the layer of neuron-rich grey matter, it's only in the last few decades that the true functionality of white matter has really come to light and we now believe amongst its many roles it plays a crucial part in learning and retaining new practical skills.

White matter is made up of bundles of fibres called axons that branch off the neurons that make up the grey matter of your brain. It's 'white' because unlike the neurons in the grey matter, the axons (nerve fibres) in this part of the brain are covered in a fatty substance called myelin. Myelin plays a crucial role in the function of these nerve fibres because it acts as an insulator allowing the passage of a nerve signal to proceed far more rapidly, jumping down an axon rather than just passing through it. Along unmyelinated fibres, impulses move continuously as waves but in myelinated fibres, they 'hop' or propagate, a process known as saltatory conduction, making them up to 300 times faster.

In this way the white matter acts as a network of bridges between areas of grey matter carrying signals between different parts of the brain. However, beyond this integrating function, we are now discovering that white matter and, in particular, the myelin sheaths also play an even more prominent role.

The importance of myelin has long been known pathologically through the study of multiple sclerosis, an autoimmune disease disease that directly damages the myelin coating of nerves in the brain and spinal cord with a resultant loss of physical and mental function. In the non-diseased brain we now know that the insulating function of myelin does far more than just speed up connections across the brain, it plays an active role in the process of learning – in fact it may well be the 'switch' that takes us from learning to learnt.

The main purpose of a myelin layer (or sheath) is to increase the speed at which impulses propagate along the myelinated fibres. Along unmyelinated fibres, impulses move continuously as waves, but, in myelinated fibres, they 'hop' or propagate by saltatory conduction.

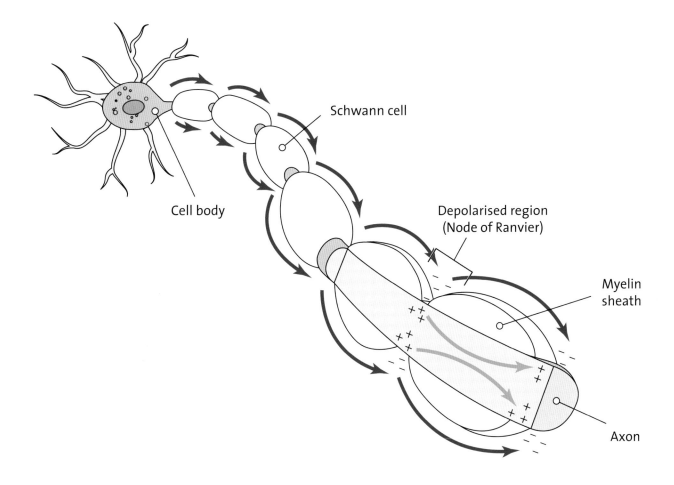

Cell body

Schwann cell

Depolarised region
(Node of Ranvier)

Myelin
sheath

Axon

Recent studies, including a series of experiments carried out at my own university, University College London (UCL), have begun to shed light on the neuronal mechanism that is behind obtaining a new skill such as riding a bike or playing an instrument. In the UCL experiments, led by Professor Bill Richardson, the team focused on the ability of mice to run on a specially adapted wheel. The wheel had its rungs spaced irregularly around its circumference, making it impossible for the mice to immediately master it – and so creating a learnt activity somewhat analogous to the challenge of humans learning to ride a bike. In the experiment two groups of mice were set the task of running on the wheel. After about two hours of practice the first group, a group of normal mice, were beginning to get up-and-running on the wheel, having begun to learn to navigate its unusual configuration. The second group of mice, however, who'd been given a drug that switched off the ability to produce new myelin, simply couldn't master the wheel no matter how long they tried for. It seemed that their learning was inhibited when the production of myelin was halted. Further evidence for the role of myelin in learning the task came in a follow-up experiment. This experiment further supported the theory that the act of myelination plays a crucial function in learning. Mice that had initially learnt to run on the wheel were subsequently given the drug to prevent further myelin production. When placed back on the wheel, the drug-treated mice were unaffected, able to immediately run on the wheel despite the ability to produce new myelin being inhibited. This suggests that once a skill is learnt, myelination is not required to remember and perform it again, it's only needed during the initial learning phase. Combining this experiment with all of our current knowledge we can now begin to describe the process that lies behind riding a bike or learning to walk.

From the first moment you begin learning to ride a bike, you are creating a network of connections in your brain unique to that activity. This 'cycling network' involves all the different components of the brain needed to balance on two wheels, from our sense of sight and touch, our balance system all integrating to the control of the multitude of muscles needed to keep you upright. At first all of these different parts of the brain are functioning independently of each other but each time we practise, each time we fall and get up again, we repeatedly fire the same neural pathways, and gradually this causes the connections between these disparate areas of the brain to strengthen and start forming a network. As this happens it triggers a type of brain cell, called an oligodendrocyte, to increase in number and begin wrapping the newly emerging network in ever thicker layers of myelin. The more often you practise, the thicker the myelin becomes, making the axons even more efficient and faster in coordinating and conveying information

around the brain. It's a process of turning a loose network into a superhighway, hard-wired into our brains, and it is this process that, once complete, creates the miraculous moment when the impossible becomes possible, when all that effort becomes effortless and we fly off on two wheels never to look back again.

This process of myelination begins early in our lives, as we learn our most important of life skills, the ones that we need to be able to do without thinking.

Walking on two feet is one of the most difficult of these and one of the first that we master; a motor skill that becomes hard-wired in our brains around the age of 18 months after many weeks of practice and failure. Once this myelinated network is in place, we never have to think about it again. However, there is a biological reason we learn so efficiently when we are young and why it feels as if learning a new skill such as playing an instrument or juggling gets harder with age. It's because the ability to generate myelin gets slower with age – we simply cannot craft a new learnt network as efficiently as in our youth. All is not lost however; new research suggests that when we learn a novel skill later in life, the brain can generate new myelin by increasing the number of myelin-producing cells. It might feel like a steeper learning curve but it's reassuring to know it might just be possible to teach an old dog new tricks.

An early (pre-bipedal) picture of Drs Chris and Xand van Tulleken.

# THE DIARY OF A HUMAN BODY AGED 37 AND THREE-QUARTERS: LEARNING TO JUGGLE

I learned this the hard way in a battle of fast hands, quick eyes and deep determination. My opponent? An 8-year-old called Tahu. The contest? Who could master juggling first … not one, not two but three balls all at once. I was confident. Tahu was nervous. Intimidated perhaps by my firm handshake, my cold stare or just the fact that a grown man was acting in a bizarre and overly competitive manner. After beautifully demonstrating the baseline for the experiment – that we were both equally useless at juggling in any capacity whatsoever – we put in about an hour's practice together. Well, I put in an hour's practice. Tahu got distracted, then embarrassed, then discouraged. I was the underdog and he knew it. But in this first hour I had a secret weapon – my pre-frontal cortex. As pre-frontal cortices go, it's not overly developed but it means that I have a capacity for pleasure delay that the average 8-year-old doesn't. And for all my bluster I was relaxed. I didn't care about the cameras (I understand that regardless of what happens, it's up to the editor whether or not I look foolish – and Tahu is in no position to buy him a beer).

At the end of this hour I had already experienced the increasingly bizarre sensation of going from conscious incompetence to unconscious competence. I could juggle three juggling scarves. I didn't quite know how but somehow my hands were able to find them in mid-air against all expectation. But it was hard graft. It's not possible to juggle in an MRI scanner but if it was, you would have seen my entire brain demanding blood. After the scarves I learnt how to throw two balls from one hand to the other, forcing my left hand to mimic my right in a way I hadn't believed possible. Then a third ball was added so I started with two balls in my right hand. Launch from the left, then the right, then, with the first balls starting to descend, launch the third ball from my right hand and moments later catch the first ball released from the left. Now the left must be ready to catch and release and this should continue in a cycle. In fact what happens is that the first release is a little too hard, so the hand is wrongly placed for the subsequent catch. This can

Juggling. Tahu about to add a third ball. He's lurching forward and will have to run about the room to continue catching the balls but he was able to sustain more than 20 catches with infuriating regularity thanks to his myelin (and his mother).

be rescued with a stumble but the mistakes amplify and pretty soon all three balls are on the floor. And as often as not, so was I. Almost immediately in fact. But I had tasted a success, a few well-caught throws and I was definitely ahead of Tahu.

Then Tahu and I both went our separate ways, with the final juggle duel just seven days away. I immediately set to the task.

Well, that's not quite accurate. I procrastinated horribly. Tahu had time on his side in two senses. His young brain is likely able to lay down new circuits and reinforce existing ones quicker than mine. But he also doesn't have a mortgage ... or not one that he mentioned. He's got a few obligations doubtless, perhaps a birthday party. But his parties won't leave him disabled for days afterwards as his liver fails to rapidly digest the toxic breakdown products of alcohol metabolism. Perhaps he has a test at school but nothing that will affect his career prospects if he fails it.

So, I squeezed in a few minutes here and there with my back aching from picking up the dropped balls. We'd been advised to juggle over a bed but, in order to give your brain time to catch up with the falling balls, you need a room with a high ceiling, and I don't sleep in a ballroom. A brief attempt at practising outside was cut short by a combination of rain, wind and the fact that the balls seemed to act as homing devices for the dog mess in my garden (I am less diligent than I thought at picking it up). Without a camera, and an 8-year-old to beat, juggling was suddenly a lot less fun. And the discovery that my wife, Dinah, is able to juggle five objects as if she had been born into the circus, was not encouraging. 'How do you do it?' I asked. 'I don't know,' she replied. 'It's easy though.'

I started to worry that my pre-frontal cortex was no substitute for Tahu's mum making him practise. She didn't strike me as the kind of person who would let her 8-year-old be beaten in a test of learning by someone three decades older.

But you don't pass as many exams as I have without knowing a few shortcuts. Yes, Tahu would have done a little more practice, but I knew that with filming comes faff. When I met him at the school I could do no more than barely catch three balls once each; little better than when I left him. But with the test looming and the camera on me, I set about practising. Something had been happening that week between the snatched practices here and there. My hands started to find the balls, I threw higher to give myself more time. By the time we were being filmed properly, I could reliably complete a full cycle of catches sometimes up to eight or nine.

It reminded me of my only other similar recent experience. Learning to present television. It is easier, certainly, than learning to juggle but I remember the main problem I had when I first started – pieces to camera. Little paragraphs delivered down the lens to the audience, to explain, contextualise, own or introduce a topic. Almost all presenters struggle with this initially. I remember needing 20 'takes' sometimes to overcome stumbles, or forgotten words. Now, almost a decade later, I can take a passage of text and memorise it in under a minute and generally get it done in a couple of takes. It's a strange feeling and it has absolutely no wider use, but years of frustrating crews with multiple takes meant that whatever circuitry or myelination is required for 'presenting television' has gradually become solidified and strengthened totally unconsciously.

Learning to juggle was a hugely empowering experience. As adults, we're not often forced or encouraged to learn new motor tasks. It reminded me that learning is not mere myelination. Yes, that is one of the neural substrates, along with making new synapses and strengthening existing ones. The practice that will drive these changes is fairly easy. The motivation, drive, time and desire to do so may be harder to come by. I have no doubt that per minute spent, Tahu got further than me, but it renewed my confidence that as we age we are still very able to learn new skills.

# TALK THE TALK

In developmental terms once you've walked the walk, it's time to talk the talk. Of all the skills that you learn in early childhood, perhaps the most miraculous is the acquisition of language. Most of us do it every day without even thinking about it, yet language is a uniquely complex human ability that each one of us seems to develop with minimal effort, at least any effort that we remember.

From the earliest babbling that starts around five to six months after birth, we begin to learn to control the creation of sound through the precise manipulation of our mouths, lips, tongue and larynx (voice box). These babbles soon begin to blend into words, but before that we go through the most frustrating of stages, a stage where we understand words but cannot produce them, a time for tantrums and sulks as the comprehension of language sits within us but the production of it remains elusive. During this phase we rely on an array of gestures (thumping fists, impatient pointing, and pouting bottom lips) to communicate our demands. However, from about 12 months onwards, language appears to invade our brains at a phenomenal rate. At first it's single, simple words, a time of 'mama' and 'dada', 'no!' and 'mine!' But then complexity slowly starts to emerge. By the time we reach 18 months, we have a vocabulary of about 50 words at our disposal and many more that we can comprehend. The progression then continues at a rapid rate, with the use of strings of words quickly followed by sentences as the size of vocabulary expands exponentially for a few years, peaking in our early twenties and then growing more slowly to

The structures of the head and neck involved in speech. The chest and diaphragm are also vital for passing the air in a coordinated fashion over the vocal cords and regulating volume.

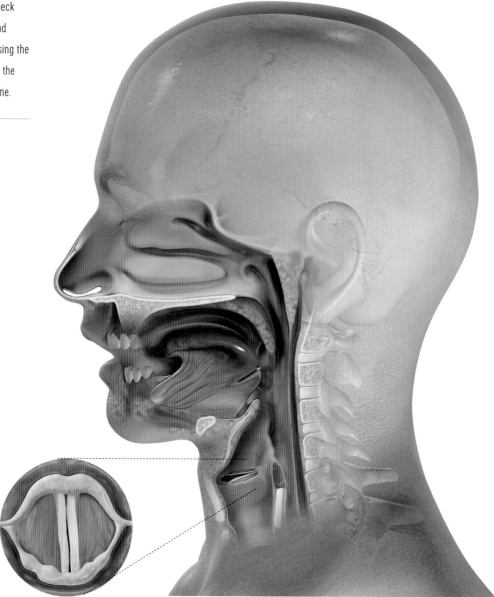

middle age when it pretty much stops. From a vocabulary of 300 words at 2 years old, to 5,000 words at 5 years old, to 15,000 words at the age of 12. The average adult human who is a native English speaker has a varying vocab depending on how you measure it. In vocabulary tests people who have not been to university will know around 35,000 words, graduates will know 50,000 and English professors upward of 75,000. Or you can look at the vocabulary of texts. Tabloid newspapers and the King James Bible both have a vocabulary of around 8,000 words. Shakespeare is often cited as having the largest vocabulary of any writer but the data don't support this. He wrote his plays with around 15–20,000 words. Impressive for the time, when a farm labourer might have communicated with only a few hundred.

If an adult vocabulary of 50,000 words isn't impressive enough, we can continue to learn up to two new words a day if we seek out the stimulus to do so. It's a process of learning that is almost miraculous to witness, but despite decades of research many of the secrets of precisely how we learn to speak remain a mystery.

# THE BIRTH OF A WORD

Technology has transformed our ability to explore human development in a multitude of ways. One of the most extraordinary uses of the emerging digital technology was employed by husband-and-wife team Deb Roy and Rupal Patel, in 2005. Back then Roy was studying robotics at MIT, whilst trying to design and build a conversational robot. Roy realised he was attempting to teach a robot to do what children seemed to do effortlessly. So it seemed a simple, if unusual, leap to want to explore childhood language development, to better understand how a robot could be taught to converse. Using the cutting-edge video technology of the time combined with emerging machine learning and pattern recognition from his robotics work, Roy had begun to record and analyse real-time language development using video and sound data captured in a specially adapted space in their faculty. This involved bringing children into the lab with their families for extended periods of time in an attempt to capture the actual process of language acquisition and development. However, the limitations of capturing real family interactions in a lab environment soon became apparent and so they began to wonder how they could capture the same type of data but in a real home environment instead. The answer came about through a beautiful collision of timings. In July 2005 Roy and Patel were expecting their first child and as a tech-savvy Dad-to-be Roy also knew that with the dawning of the age of big data, the far-fetched idea of recording every moment of his child's first few years of life had gone from science fiction to fact. And so began the Human Speechome Project: rigging their home with a specially designed video and sound system, cameras and microphones recorded every second of their son's life through every waking hour. Creating the largest home video collection in history, the project amassed over 200,000 hours of recordings across their son's first two years of life, capturing over 80 per cent

The Human Speechome Project: sample video image from the kitchen.

of his waking hours. From the earliest babble, to the emergence of his first word, contained within these 16 million recorded words are the secrets of how we learn language. Just watching and listening to the birth of a word like 'water' emerging slowly over days and weeks until it is fully formed and utilised reveals a deeply profound and human process. Roy and his team called these moments word births, word histories that could be tracked and traced across time and space.

Picking their way through this vast amount of data, Roy and his team also began to see patterns emerge that were at times expected, at times surprising. A vocabulary burst around the age of 1 is a common occurrence and this is exactly what the footage demonstrated, but less expected was a sudden crash in the learning of new words around 19 months. Intriguingly, further analysis showed that at exactly the same time as this dip, the production of two-word sentences took off, suggesting that his brain switched from word acquisition to grammar acquisition.

With such a vast amount of data to sift through, the Speechome project remains a treasure trove of research opportunity and could potentially continue to yield insights for years to come. For Deb Roy and his family it will remain an extraordinary, perhaps unique, record of his child's first two years of life. It also led Roy to use the algorithms and pattern recognition systems he innovated for the project for a very different purpose. Roy's Bluephin lab uses the tech to monitor the vast streams of social media conversation around TV shows and ads and looks for patterns in the data that are commercially valuable. Being able to listen carefully really is a skill that pays off.

## UNIVERSAL GRAMMAR

Scientists and philosophers have long tried to understand the nature of human language acquisition, grappling with the basic question of nature versus nurture. Is language a learnt skill, absorbed from our environment and laid onto the *tabula rasa*, the blank slate of our brains throughout our early years? Or is it an innate function of the human brain, driven by evolution through natural selection, initiated like all advantageous characteristics by the chance genetic mutation in our ancestors? The exploration of this question has been dominated in the last 50 years by the work of Noam Chomsky. Chomsky is often held up as the father of modern linguistics, transforming the study of language from semiotics to science. Since the 1960s he has driven the idea that we are born with an innate linguistic ability, that language becomes hard-wired into the human

Wordscape showing the birth of a word by mapping the data related to every utterance of the word 'water' in Deb Roy's home. (Philip DeCamp/Deb Roy)

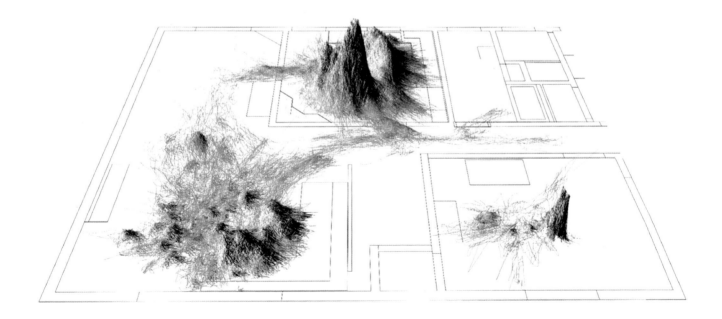

brain, a 'universal grammar' that we all share at a biological level and that links all human languages that rely on this intrinsic ability. Chomsky controversially argues that it was one chance mutation in an individual around 100,000 years ago that gave rise to our linguistic ability in one giant leap. Although this is far from the consensus thinking today, Chomsky's wider theories continue to drive the search for evidence to support the idea that language is innate.

One intriguing piece of research exploring the origin of language in the brain has emerged from an unlikely source. Playing music to babies may not seem like the most interesting way to get to the root of the human language problem but Dr Reyna Gordon and her team at Vanderbilt University Music Cognition Lab have begun to reveal that the very basis of language is underpinned by something unexpected – rhythm.

Before we explain a little bit more about Dr Gordon's research, I just want you to think for a minute about a strange feeling I'm sure most of us have had at some time. It's that moment when you're in the car or kitchen, listening to the radio and a tune pops up that you literally haven't heard for years, maybe even decades. Suddenly out of nowhere you're singing along word perfect (if not quite pitch perfect in my case), performing a seemingly miraculous feat of memory by recalling all of the lyrics to this long-forgotten song. You probably won't remember the name of the song, or even the singer, or the band, but something in you finds the ability to recall the lyrics. Why is that? What is it about music that enables our memories to act in such a surprising way? The answer it seems goes back to an innate relationship we all have with rhythm.

When Dr Gordon gathers a group of babies in her lab in Nashville, Tennessee, and plays music to them with a range of different tempos, it's clear they are all trying to dance in time – some better than others. Matching a beat is an instinctive reaction to music, an innate ability locked into our brains at birth that we believe is universal. Get it right and the babies exhibit a clear sense of pleasure; their limbs might not be moving perfectly in time to the beat but it's clear the changes in music tempo elicit changes in body movement.

Not only can we match a beat, we also have an ability to predict what's going to come next. We are able to respond to rhythmic changes both predictively and flexibly, an ability that is known as phase alignment. Even our closest cousins the chimpanzees are not able to do this; it's a purely human ability, as is the ability for spoken language. And we now think that being able to sense and predict rhythm is an ability that is closely connected to our brain's ability to process language.

From the moment our auditory cortex starts to function at 24 weeks through gestation, the first thing we hear is the rhythm of our mother's heartbeat. This introduction to rhythm, even before we hear individual words, appears to be a key moment in how we learn language. The reason is that it prepares our brains for the complexity of spoken language, ultimately enhancing our ability to understand what's being said by helping our brains predict what's coming next. By reading the rhythm contained within language, it allows us to predict the most likely scenario and so language becomes effortless, automatic. It's why we can remember the words to a song we haven't heard for years.

Across a range of different studies Reyna has shown strong correlations between rhythm skills and language skills. We now know that those children who are better at matching a beat go on to be better talkers at the age of 6, because rhythm shapes the way we use and understand language. It also seems that children with language difficulties are helped by improving their rhythm skills through the simple process of learning a musical instrument.

This research adds another little piece to the puzzle of human language acquisition and it also strengthens the idea that language is an innate part of being human, a skill that we are all born with. But to turn that innate skill into a fully functioning grasp of language, each one of us goes through a process of learning that requires the building of a vast vocabulary of words. A process that takes us to another mysterious part of the functioning of our brains – the secret world of memory and in particular how we move the filing cabinets of our minds between short- and long-term memory.

# THE LONG AND SHORT OF IT ...

If I gave you a list of ten words right now, words that you've probably never seen before like *abomasum* (the fourth stomach of a ruminant such as a cow or sheep) or *ecdysiast* (a striptease performer) or even *sesquipedalian* (a word of many syllables or a long-winded piece of writing), how quickly could you commit them to memory? And what would that memory actually be in your brain? Would you be able to remember the word, spell it and retrieve its meaning in an hour's time or tomorrow or even this time next year? And what would that process of memorising involve? The secret of how our memories function is something that has fascinated and beguiled us for centuries. In the past it was left in the realms of philosophy and education to attempt to gain an understanding of the basis of

**TEN UNUSUAL WORDS**

- apoptosis
- entomophagy
- jumentous
- borborygmus
- portolan
- strappado
- inspissate
- eucatastrophe
- sesquipedalian
- kinnikinnick

Akash Vukoti, at 2 years old, during his first spelling competition, the MastiSpell competition, 2012.

memory. The method of loci, or memory palace technique, is centuries old and involves placing objects in an imagined three-dimensional space – a palace of the mind. The technique is attributed to the ancient Greek poet Simonides, who realised, after a lucky escape from a banqueting hall which collapsed, that he could remember the names of his fellow guests by picturing the places they were sitting. This technique enabled him to identify mangled bodies crushed beyond recognition. The technique has been developed now to the point where after immense practice, memory champions like Suresh Kumar Sharma can memorise Pi, an irrational number whose decimal representation never ends and never repeats, to over 70,000 decimal places.

'THE LONGEST WORD I CAN REMEMBER IS PNEUMONOULTRAMICROSCOPICSILICOVOLCANO-CONIOSIS ... IT'S A KIND OF LUNG DISEASE.'

**AKASH VUKOTI, 7 YEARS OLD**

But while mnemonists over the centuries have been driven to astounding feats of memory, research across the last century has only recently opened up the black box of our brains and for the first time allowed us to begin to understand the mechanism underlying a deceptively simple process like learning a new word.

During childhood we learn new words at an extraordinarily rapid rate. The dictionary stretches to around 300 words for a 2-year-old and then we start adding around a thousand words a year or so as we move through childhood. So, on average, by the age of 7 we are playing with around 6,000 words – learning about several new words a day. But Akash Vukoti is no average 7-year-old. For the last three years he has been competing in the viciously competitive world of spelling bees. With a vocabulary of 50,000 words, he's more akin to a 12-year-old in his spelling ability, a precocious talent that has resulted in him becoming the youngest participant in the finals of the most famous competition in the US, the Scripps National Spelling Bee. Akash spends two hours a day with his older sister Emrita memorising new words.

π to 1,001 places.

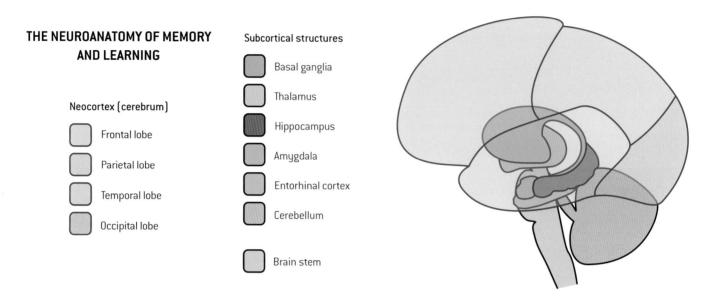

**THE NEUROANATOMY OF MEMORY AND LEARNING**

Neocortex (cerebrum)

Frontal lobe

Parietal lobe

Temporal lobe

Occipital lobe

Subcortical structures

Basal ganglia

Thalamus

Hippocampus

Amygdala

Entorhinal cortex

Cerebellum

Brain stem

It's a process that is familiar yet elusive; we instinctively know how to learn something through repetition but what is the physical process that allows a new word to become permanently stored? When we try to learn new words, we are turning experience into something permanent and physical that can be recalled. This process of creating a short-term memory requires the outside to be brought in and physically incorporated into our brains.

Classic cognitive psychology breaks down the process of memory into three key stages – encoding, storage and retrieval. The encoding stage is all about sensory input entering the brain as predominantly visual, acoustic or semantic information. When Akash is learning a new word, he is using all three of these types of input. He sees the new word, hears it by saying it out loud and understands it by learning the meaning of the word. These sensory impulses activate different and specific brain regions – the visual cortex at the back of the brain, the auditory cortex in the temporal lobes – and as he learns the meaning of the word the anterior temporal lobe becomes activated. When we are presented with a list of new things to remember, we instinctively attempt to learn them through the auditory route, repeating or rehearsing the information by saying it out loud again and again, but it's the combination of inputs that is the key to creating a new memory.

The association between the different stimulated brain regions is the foundation of storing any new memory. It's only in the last few years that scientists have revealed the physical secrets of a memory actually being formed.

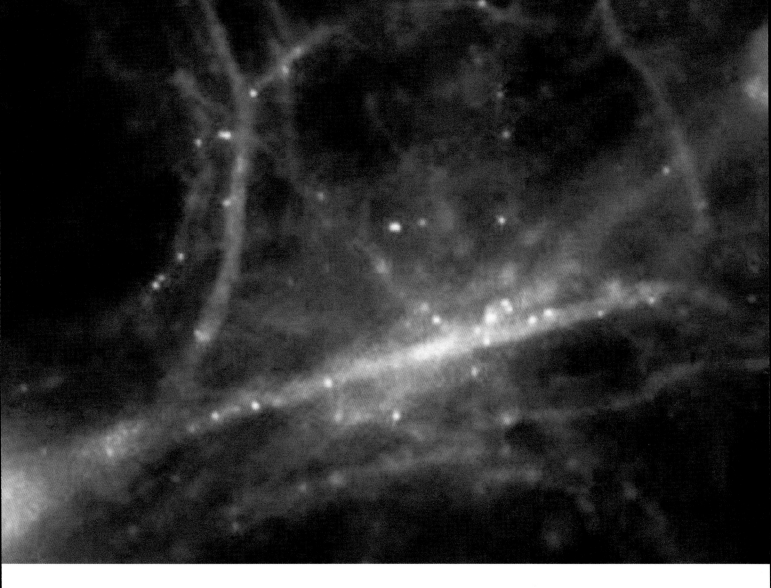

This extraordinary image shows the actual physical process of memory formation. Each one of the bright spots in the image is a fluorescently tagged molecule called a messenger RNA (mRNA) molecule and it's the activity of these molecules travelling down the neuronal branches of your brain that we now think is the key to how we make new memories.

For an experience to transform into a memory, the network of stimulation sparked by the sensory input needs to be preserved in the neuronal structures of

Molecules morphing into memories in the brain – ground-breaking work by researchers at the Albert Einstein College of Medicine of Yeshiva University, New York.

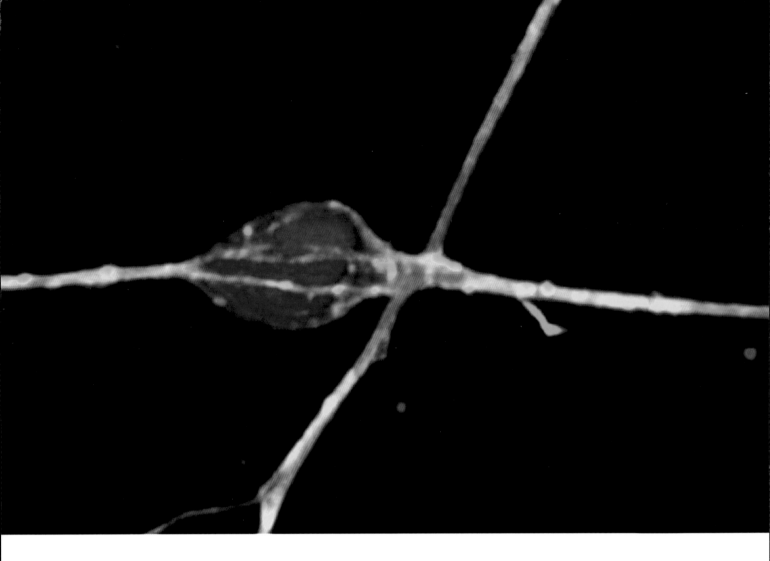

A migrating cerebellar granule neuron in culture showing the nucleus in blue, microtubules in red and the actin in green.

the brain and the synapses that connect them. In the case of learning a new word such as *abomasum*, *ecdysiast* or *sesquipedalian*, this means the auditory, visual and semantic input needs to be captured. The key structure in the brain that enables this to happen is called the hippocampus – named after its resemblance to a seahorse. Part of the limbic system, the hippocampus is found deep in the brain. Each one of us has two hippocampi, one located in each side of the brain and it's these structures that play the crucial coordinating role in consolidating a memory. When the different regions of the cortex are stimulated by learning a new word, this activity is mapped onto corresponding regions of the hippocampus and it's here that the memory consolidation process is initiated and controlled.

Recent research at the Albert Einstein College of Medicine has begun to reveal for the first time the precise mechanism that occurs when a new memory is stored in the brain. The secret to building a new memory starts when neurons form stable, long-lasting contacts between each other. Neurons in general are long and thin with a bulge where the nucleus is. At their ends they touch other neurons with long finger-like spines called dendritic spines. A synapse is where these spines contact the spines of other neurons and pass chemical messages from cell to cell in the form of neurotransmitters. Memory and learning involves the strengthening of these synapses through repeated stimulation, and a protein called beta-actin appears to be crucial for this. Beta-actin is found in all cells and has a vast range of functions – it is a vital part of the cytoskeleton, the protein fibres that give cells strength, shape and integrity. It is closely related to the protein actin found in muscle cells that allows your muscles to move. Crucially beta-actin helps cells move and change shape, which is vital in the developing embryo. It may be this function which makes it so central in making memories. When a memory is created beta-actin appears to physically change the shape of the dendritic spines. The question for the team at Einstein was how is such a common protein, expressed in all cells, regulated to create the precise structures required for memory formation?

The answer starts deep within the hippocampus in the nuclei of specific neuronal cells. It's here that molecules of mRNA for beta-actin are manufactured. For any protein to be made, the DNA code in the nucleus of a cell is first turned into a very similar molecule to DNA called mRNA. This mRNA in turn codes for the sequence of amino acids which makes a protein. In this case it is the mRNA molecules that are tasked with actually building the memory. The crucial thing is not the protein being made – beta-actin is after all a ubiquitous protein – but the exact point at which the mRNA is translated into the beta-actin.

In work carried out by the team at the Albert Einstein College of Medicine, they used mice neurons to track this process in extraordinary detail. When the mouse hippocampal neuron is stimulated, within 15 minutes mRNA molecules coding for beta-actin are made in the nucleus. Using a fluorescent dye they were able to tag and track mRNA as it was being formed in the nuclei. They then watched as this mRNA left the proximity of the nucleus, travelling through the nerve cell down the dendrites (branches) until the mRNA reached the end of its journey at one of the billions of synapses (junctions) that link neurons together.

The crucial discovery was how then the beta-actin is deposited at the correct synapse. Intriguingly it seems brain neurons are unique amongst nerve cells in that they control not only the production of beta-actin mRNA but also the moment at which it can be turned into protein. mRNA is packaged into granules until it can complete its journey through the neuron and reach its intended site of activity. Stimulating the neurons disintegrated the granules and allowed the protein to be made at precisely the correct site. A few minutes after stimulation the granules reform and the mRNA is packed once again, preventing too much beta-actin from forming. 'Frequent stimulation of the neuron would make mRNA available in frequent, controlled bursts, causing beta-actin protein to accumulate precisely

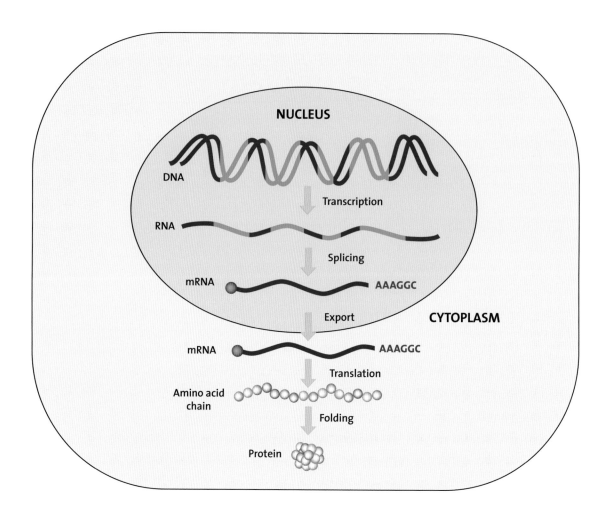

**NUCLEUS**

DNA

Transcription

RNA

Splicing

mRNA • AAAGGC

Export **CYTOPLASM**

mRNA • AAAGGC

Translation

Amino acid chain

Folding

Protein

The Central Dogma of Molecular Biology. You are made of protein. This is how proteins are created from DNA, which is first turned into a chemically similar intermediate called mRNA, in a process called transcription. This is exported from the nucleus and then 'read' by proteins called ribosomes in a process called translation. This creates the amino acid chain that folds up to form a protein.

where it's needed to strengthen the synapse,' said Dr Singer, head of the lab in which the research was done.

It's thought that the strengthening of these synaptic connections is the key to making a new memory and we now know that beta-actin is a crucial protein in altering the shape of the dendritic spines and strengthening these connections. When the tagged mRNA arrives at a synaptic terminal, it begins to synthesise the new proteins allowing the neuron to generate these brand-new connections, essentially preserving the memory in the anatomy of the brain. In this way, by building and strengthening the connections between neurons, an ephemeral short-term memory is transformed into a long-term memory.

The deeply layered beauty of this research is astounding. Years were spent by one of the lead researchers developing a system where mRNA molecules could be given a fluorescent tag without disrupting their function. This allowed them to be visualised and enabled understanding of the formation of the granules. But even

■

'FREQUENT STIMULATION OF THE NEURON MAKES mRNA AVAILABLE IN FREQUENT, CONTROLLED BURSTS, CAUSING BETA-ACTIN PROTEIN TO ACCUMULATE PRECISELY WHERE IT'S NEEDED TO STRENGTHEN THE SYNAPSE.'

**DR ROBERT SINGER**

■

as we gain an insight into one of the molecular bases for memory, this research shows that we are still learning about control of one of the most basic principles of molecular biology, the transformation of DNA to RNA to protein. The team at Albert Einstein show how intricate the mechanisms are that enable our simple four-letter DNA code to create one of the most fundamental parts of any human being – our memories.

But according to the latest research, if we really want those newly formed memories to stick, we need to stop trying to learn that new word and give our brains a chance to consolidate all the information by going to sleep. According to some of the latest sleep research, during periods of slow wave sleep (SWS) we reactivate recently encoded memory networks. They are then shifted to more efficient and permanent brain regions, making things easy to recall the next day.

So when you wake up tomorrow ask yourself if you can remember the name of a cow's fourth stomach, or the correct way to describe a striptease performer and if you have really strengthened those synapses with plenty of beta-actin you may even be able to recall the word that describes a word of many syllables.

This type of memory – the ability to recall facts and events – is often referred to as semantic memory. It is the repository for our general knowledge of the world and is of course crucial to the functioning of our everyday lives. It was first defined in the 1970s by eminent Canadian neuroscientist Endel Tulving (see p. 141) and we still often use this classification today. He divided memory into two main categories – implicit memory and explicit memory. Implicit or procedural memory refers to the memory of skills and abilities like riding a bike, driving a car or playing the piano, memories that we recall entirely subconsciously every time we repeat the activity. It's why, as we've seen earlier in the chapter, we don't have to go through a conscious process of recall when we jump onto a bike or get into a car. Explicit or declarative memory on the other hand requires a far more conscious act of recollection. Tulving split this type of memory into two further subcategories – semantic and episodic memory. The semantic memory, as we have seen, is the memory of words and facts, whereas episodic memory is the autobiographical part of our memory, the precious place where we store our experiences. These memories of events, interactions and moments both good and bad are entwined with emotion and lie at the very heart of our identities and personalities. They are in many ways the memories that make us who we are and it's this that we turn to next.

# MEMORIES ARE MADE OF THIS

What is the oldest memory you can recollect? The moment furthest back in time that you can retrieve from the deepest vaults of your mind? For me it is a moment from the summer of 1981.

I'm just 3 years old. We are swimming at a pool in London. The pool has a wave machine. Xand thinks it will be funny to push my head underwater. He's half right. It's hysterically funny for him and very unfunny for me. I remember my rage and tears. I remember vowing that one day I would take revenge on him even while I coughed up water that, typically for a 1980s municipal pool, was full of matted hair and plasters.

There are a number of features of this that make it a good candidate for a first memory. The fear and anger. The re-visitation of it over the following days as I plotted revenge (as I write I realise that this particular crime is unavenged – I must take Xand swimming). Is it actually a memory or just the layers of a family story told and retold over the years until it becomes belief more than recollection? Or is it truly a moment captured by biology, a moment when my brain switched from letting the world fly through it, to allowing the world to stick, leaving a chemical, structural representation of the event in my neurons?

Nobody as yet fully understands why we all exhibit something commonly known as childhood amnesia. Universally, it seems, we are unable to recollect a specific event before the age of 2 at the very earliest and most of us carry very few episodic memories – our memories of places, people and things – from the first

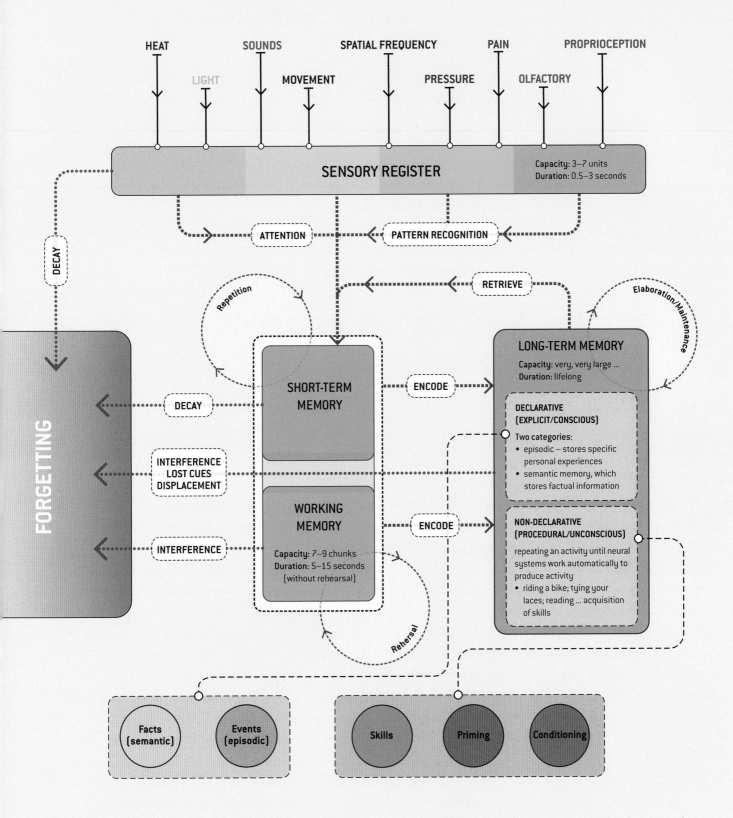

Conscious Learning: declarative memory will naturally decay in an exponential way over time and be lost to recall. Repeated review of the facts or episodic details at decreasing intervals restores the information and subsequent decay is slower and shallower. Eventually the information will be permanently stored, with relatively minimal decay.

Molaison's brain was frozen cryogenically into a block of gelatine and scanned to a 3-D model using a series of digital images of the block's surface. These were obtained using a digital camera mounted directly above a microtome stage – 2,401 digital anatomical images and selected corresponding histological sections that were collected over the course of an uninterrupted 53-hour procedure.

decade of our lives. Yet this is the decade through which we have some of our most intense experiences as we encounter the world for the very first time.

Understanding the reasons for childhood amnesia is still beyond us. And we still don't understand the physical mechanism behind what allows our brains to record something at the age of 3 that will still be accessible to you even a century later – the connections and proteins which mean that at the end of our lives we can see, hear, and feel a few moments from the first few years. But the recent explosion in neuroscience has allowed us to begin unpicking the details of how an autobiographical memory is laid down.

Much of this understanding began to emerge in the last century by studying the behaviour of people who had survived serious brain damage. Throughout the twentieth century advances in trauma medicine and antibiotics combined to transform the survival rates of patients after head injury and stroke. At the same time surgeons were also pushing the boundaries, although often with ill effect, allowing us to gain specific insights into the function of different parts of the brain. One of the most famous and revealing cases of this type was the strange story of a man called Henry Molaison.

# CASE NOTES: HENRY MOLAISON – THE MAN WHO FORGOT HIMSELF

Henry Molaison was born in Manchester, Connecticut, in 1926 and died 82 years later just a few miles upstate from his place of birth. His was a generation that lived through an endless stream of extraordinary events, from global war to seeing a man walk on the moon, and yet Molaison experienced these events in a way that intrigued and puzzled neuroscientists for decades. From the age of 10 Molaison suffered from increasingly severe epileptic seizures. Triggered, perhaps, by a bicycle accident a couple of years earlier, by his late teens he was struggling with the severity and frequency of the seizures as they impacted more and more on his ability to lead a normal life. With high doses of anti-convulsant medication having little effect, he was eventually referred to the care of William Scoville, a neurosurgeon at Hartford Hospital, Conneticut. Investigations by Scoville led him to conclude that the epilepsy was localised to the temporal lobes of Molaison's brain and, as was the growing surgical fashion at the time, the decision was taken to perform an experimental procedure known as a bilateral medial temporal lobe resection – in other words they would attempt to control the epilepsy by cutting out a substantial part of Henry's brain.

On 1 September 1953 Scoville opened up Molaison's skull and dissected out several deep-seated structures on both sides of his brain, including his hippocampi, amygdala and entorhinal cortex. On awakening, the primary goal of the surgery appeared to have been successful. Molaison's epilepsy had been abated by the operation but the side effects of having such fundamental parts of his brain removed were devastating. From the time he woke up in the recovery room of Hartford Hospital, it was as if Henry Molaison's memory was frozen in time. Unable to commit any new information to his long-term memory, the events of the day were forgotten almost as fast as they occurred. Events before the surgery remained vivid and real in his memory but beyond that point the filing cabinet of his mind appeared to be locked from receiving any new files. According to the early reports, Molaison underestimated his own age, forgot meeting people he had been introduced to minutes before and yet was able to conduct a normal conversation and

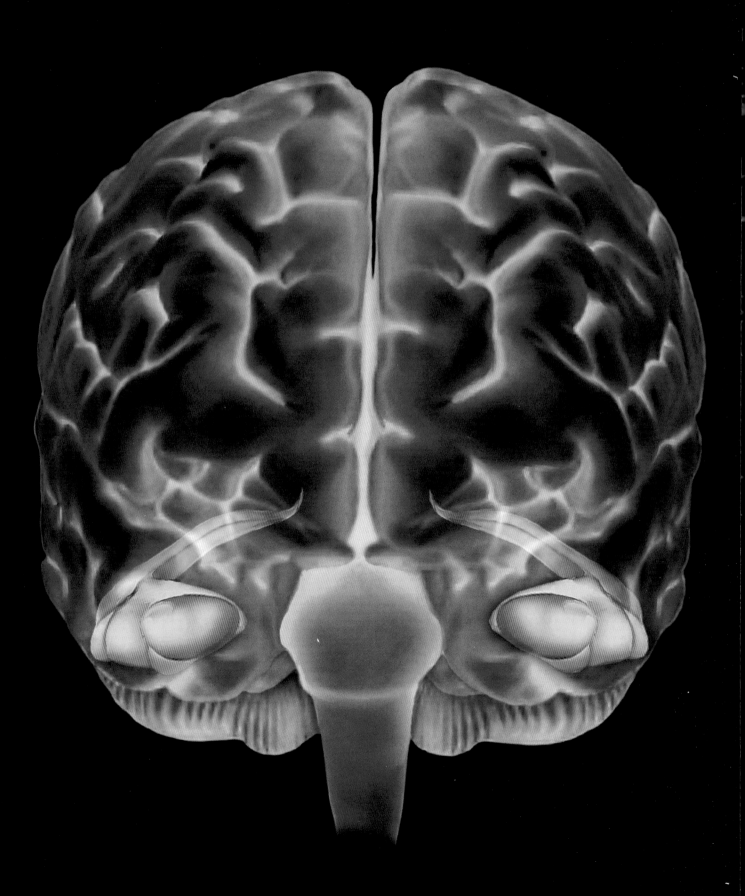

even retain a string of numbers. But nothing he would experience would find a place in his memory beyond 15 minutes or so of recall. For the next six decades he would watch the twentieth century pass by him – to Molaison the world had been frozen in 1953. Perhaps even more distressing was the fact that his cognitive and intellectual abilities were not further affected by the operation and so, in a way almost inconceivable to anyone but himself, he was painfully aware of the deficit the surgery had left him with. 'It's like waking from a dream … every day is alone in itself …' he said when asked to describe his own state to Brenda Milner, one of the early investigators who first studied him. That dialogue with Milner and the wider scientific community would go on for the rest of his life and beyond, because for all of the tragedy of his condition, the specifics of his deficit would make him one of the most important human research subjects of all time, revolutionising our understanding of how memory works.

Over the next 50 years the detailed study of Molaison and his memory deficit would help establish many of the fundamental principles of how we believe memory function to be organised in the brain. If the ability to store new long-term memories could be lost whilst still retaining the memories that were formed before the surgery, it suggested that the long-term storage system sat outside of the areas like the hippocampi that were removed in the operation. Careful investigation of the actual damage caused to Molaison's brain in the following years, including the use of MRI much later in his life, revealed the detailed anatomy that appears to lie beneath these two separate systems. Using the findings from Molaison (and other patients with localised brain lesions), combined with laboratory work that attempted to mimic his impairment in animal models, it has been possible to create a detailed map of the structure and function of what is now termed the medial temporal lobe memory system. This network of structures in the brain connects the hippocampi, which as we have already seen in this chapter are crucial to memory function, with a series of adjacent structures including the entorhinal, perirhinal and parahippocampal cortex. For the last few decades this system has been at the heart of our understanding of how a memory transforms from short term to long term as these structures bind together the diffuse range of storage sites in the rest of the neo-cortex to create the representation of a whole memory. Thousands of studies have backed up the evidence accumulated from the initial studies of Molaison and created an enduring theoretical basis for our understanding of memory. At the heart of this is the idea that there is a distinct timeline for the formation of a long-term memory that starts in the hippocampus as a short-term or working memory and is then gradually transformed into a long-term memory by a process that transfers the storage of the memory to the pre-frontal cortex.

**Opposite page:** Coloured 3-D MRI scan in frontal view of the healthy brain of a 30-year-old patient. The two hippocampi (pink) are visible in the temporal lobes, either side of the brainstem. The hippocampus plays important roles in memory and in navigation.

**Opposite page:** Neurons in the hippocampus of a brainbow transgenic mouse.

Clive Wearing's diary. Wearing is another famous patient with chronic anterograde and retrograde amnesia. He was a musicologist, conductor, tenor, keyboardist and expert in early music when he contracted a herpes virus encephalitis which damaged his hippocampus. Since this point, he has been unable to store new memories. Page after page of his diary is filled with similar entries.

However, recently a ground-breaking piece of research conducted by a team led by Professor Susumu Tonegawa at the RIKEN-MIT centre for Neural Circuit Genetics in Japan has challenged the conventional model of how memories are formed and suggests a very different mechanism and timeline. Using a novel method, called optogenetics, the team of scientists were able to activate specific cells in the brain, enabling them to trigger a specific memory. Using this incredibly accurate method of memory control in a mouse model, they demonstrated that the process of conversion from short- to long-term memory is far from the linear one we believed it to be. Instead they found detailed and direct evidence that suggests short- and long-term memories are laid down at exactly the same time in different parts of the brain. The work suggests that when a new memory is first formed, it is stored in parallel, both in the hippocampus as we would expect for short-term memory retention but also in the pre-frontal cortex. The brain, it seems, creates two versions of a memory at the same time, but these memories are not created equally. Whereas the hippocampal short-term memory is immediately strong and vibrant, the memory laid down in the pre-frontal cortex is 'silent', taking another two more weeks to mature into a fully recollectable event. At the same time the

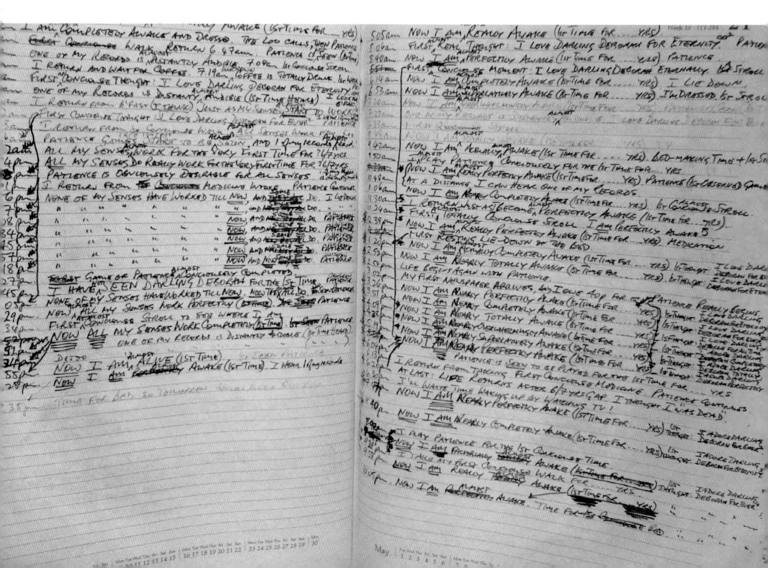

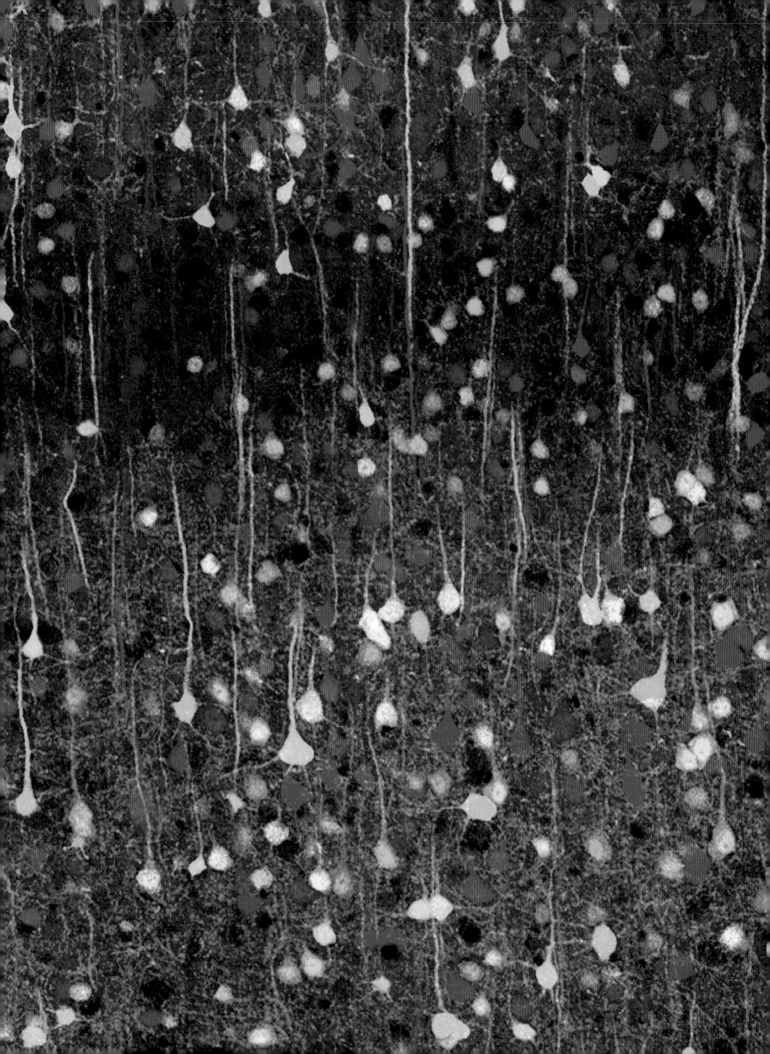

strength of the short-term memory appears to fade away – not disappearing completely but not needed for the recall of the event. It's as if the brain takes two identical images of the same event, one that is immediately in full resolution and vibrant but rapidly begins to fade, the other a much hazier image that will take two weeks to fully develop. 'It challenges the notion that there's a movement of the memory trace from the hippocampus to the cortex, and makes the point that these circuits are engaged together at the same time,' says Paul Frankland, one of the principal investigators in the study. 'As the memories age, there's a shift in the balance of which circuit is engaged as a memory is recalled.'

It's an intriguing study that really does seem to be opening up a new understanding of memory systems and the secrets that lie behind such a profound part of our identity. But it also shows just how little we truly understand about memory formation and how far we have to go until all of its secrets are revealed.

The process of learning that takes us from the profound helplessness of a human being at birth to a highly skilled individual with complex motor, language and intellectual capabilities is still shrouded in mystery and full of secrets. The explosion of neuroscience research in recent years has propelled our understanding of the process of learning in an array of ever increasing and surprising directions. From the crucial role of networks of myelin in getting us up on two feet or two wheels, to the significance of rhythm in getting us talking. We've also seen the very process by which new memories are formed and perhaps even glimpsed the process by which those precious memories are taken and stored for a lifetime. In our exploration of the minute details of the human body, we have also discovered that it's not just the brain that 'learns' but our bodies as well.

# LEARN – BODY

On 22 March 1995 Russian cosmonaut Valeri Polyakov returned to earth aboard a Soyuz capsule, plummeting back to the ground at up to 17,500 mph before landing safely in the flatlands of the Kazakhstan steppe. Emerging determinedly from the charred space capsule, he walked unaided the few small steps to the awaiting deck-chair that greets, rather unglamorously, every space-farer who returns on the Russian-built craft. As Polyakov walked those few short steps, he was breaking a strict part of Roscosmos protocol: astronauts are so weak after their time in space and so battered by the ballistic return to earth on the Soyuz, they are normally carried out of the craft and placed like newly born babies into the waiting chair for the press photos and few words that mark every return. But Polyakov had a big point to prove: he had just completed the longest duration space mission of any human in history, a record that still stands to this day. Having spent 437 days aboard the Mir space station he was determined to prove that perhaps one day a human would endure an equally long space flight before stepping out not onto earth but onto the surface of another planet.

During his 7,000 earth orbits, Polyakov completed the longest space flight in history and also completed the most exhaustive test of the human body's ability

to deal with the hostile environment of space. Every astronaut who spends an extended period of time in the microgravity of a space station faces a raft of physiological changes as the body attempts to adapt to the unearthly habitat.

The physiological effects of extended microgravity have become of increasing interest in recent years as the prospect of a manned two-and-a-half-year round trip to Mars slowly comes into view. More recently American astronaut Scott Kelly spent almost a year in space whilst his astronaut twin brother acted as a control back on earth to compare the changes that occurred to Kelly during his time on the International Space Station (ISS). All of the data from this and a growing number of long-term studies, including the monitoring of Polyakov's body for two years after his return to earth, has revealed a consistent range of adaptations that happen to the body in space.

The most immediate effects of weightlessness are on the balance system, with the disorientation creating a condition akin to being carsick known as space sickness. The redistribution of fluids in the body also rapidly causes the sinuses to clog up and the face to become puffy. Eyesight is also affected, with many astronauts developing a need for glasses to help them see close up. The reason seems to be that the eyeball becomes a little squashed and the optic nerve a

Cosmonaut Valeri Polyakov looks out the window of the Russian space station Mir during its rendezvous with the space shuttle Voyager, 6 February 1995.

Expedition 35 Commander Chris Hadfield (left), Russian Flight Engineer Roman Romanenko (centre) and NASA Flight Engineer Tom Marshburn sit in chairs just minutes after they landed in a remote area in Kazakhstan after five months onboard the International Space Station.

little swollen but nobody knows exactly why; the only solution is to keep the ISS well supplied with reading glasses. However, beyond these slightly curious changes are a series of serious physiological and anatomical transformations that profoundly affect the health of astronauts. In microgravity, muscles atrophy quickly, as the fibres shrink with the enforced inactivity of weightlessness. And because the bones don't need to support muscles in the same way and so are not exposed to the same forces, astronauts can lose a fifth of their bone mass across a long mission. The effects are also far reaching on the cardiovascular system, as the weightless environment reduces its workload, with multiple studies showing the heart can reduce in size and change shape. Despite all the interactions that are put in place to counter these effects, it can take the bodies of long-duration astronauts two to three years to recover from most of them, but some will last a lifetime. As Commander Chris Hadfield so powerfully described, the challenges on returning to earth reveal just how quickly and overtly the body adapts to a new environment.

Space travel may be the most extreme example of how our bodies 'learn' from an environment and adapt to it, but back here on earth the physiology and anatomy of every one of us is etched with the direct influence of our surroundings and the activities we partake in. As will see in the following chapter ('Survive'), the human body is designed to maintain a variety of internal parameters such as temperature, glucose levels, pH etc that can only differ across a tiny range of levels. These systems are acutely connected to the environment on earth. We live on a planet with 20 per cent oxygen with a gravitational pull of 10 m/s², and so our bodies are tuned to these conditions and change rapidly when the conditions change. But beyond the second-by-second strict homeostatic demands, over the longer term the things we do every day here on earth can also have a big effect on our bodies.

In the making of the television series, we featured Freya Christie, an up-and-coming tennis player who at the age of 19 is just beginning to make her mark on the intense competition of the women's professional circuit. Freya has been playing tennis since she was 5½ years old and during that time her body has learnt to cope with the demands she's placed on it, in a surprisingly specific way. All sport played with the intensity and frequency of an elite professional will have dramatic effects on the human body as it learns to adapt to the increasing demands placed on it, but what's particularly interesting about tennis is that it very much divides the body in two – it's an asymmetrical activity and that distinction is reflected inside as well as out. Whether serving at 100 kph or receiving forehands down the line, the forces, loads and motions of Freya's right arm, her racket arm,

cause the muscles in that arm to squeeze, squash, bend and twist, and that exposes the bones in Freya's racket arm to a huge set of repeated forces that have slowly shaped her body over the last 15 years.

Just like an astronaut being exposed to the hostile environment of space, so Freya's body has been exposed to the bespoke environmental forces of professional tennis. Whilst from the outside the two sides of her body might not look any different, inside the bones of her right arm are 40 per cent thicker and denser than in her left. At joints like the wrist, this is caused by an increase in bone size and density because the joints take the brunt of the loading, whilst in the shaft of her bones this increase results just from greater bone size, due to extra bone being laid down on both the inner and outer surfaces of the bone, but little difference in density.

In this way Freya's skeleton has learnt to cope with her intense daily activities by adapting its shape and form, differences that are more pronounced than if she was a fully grown adult because her bones are still actively growing. During adolescence the sex steroids oestrogen and testosterone promote growth by stim-

Freya Christie plays a backhand during day one of qualifying of the Aegon Open at Nottingham Tennis Centre on 10 June 2017, Nottingham, England.

ulating the secretion of IGF-1, which adds bone to the length and width of long bones like those in Freya's arms.

In fact, although Freya's body displays extreme changes due to her physical exploits – the differences in tennis players are about 10 times greater than in non-players – we all have small differences in the size of the bones in our arms, because the vast majority of us tend to favour one arm over the other when writing or carrying a bag.

However, it's not just Freya's bones that have changed, the cavities in her heart have become larger, allowing more blood to be pumped around her body in a single beat – her resting heart rate is just 47 beats per minute (compared to an average of 70 in other 19-year-olds) and an increased blood volume means her body becomes more capable of dealing with the demands of endurance exercise.

The structures in her joints have also adapted. During a game, the tendons in her elbow, shoulder and wrist are pivotal in transmitting force from muscle to bone, allowing the precise, controlled and strong movements she needs. Freya's racket arm tendons have become stiffer, transferring these forces more efficiently, and as her tendons strengthen, so too do her muscles.

Skeletal muscle is perhaps the most adaptable tissue in the human body. Made up of muscle fibres, each of these fibres contains a bundle of smaller myofibrils and each myofibril consists of a thin and a thick myofilament. These are arranged in individual units – sarcomeres – the fundamental components responsible for muscle contraction. When muscles are worked hard during one of Freya's on-court training sessions, these muscle fibres become injured. Satellite cells – specialised stem cells that generate new muscle cells – which normally lie dormant around the muscle fibres, activate, proliferate and attempt to repair the damage, but during regeneration, these cells fuse to existing muscle fibres or fuse together to form new myofibres, increasing the size of our muscles. It's a strangely counterintuitive idea but building your muscles requires you to damage them first.

All of these changes that her body illustrates combine to form a striking example of how nurture changes nature. It's something that each one of us carries within our bodies, we are adaptive machines that display the demands we put on our bodies in the intricate individual detail of our own anatomy and physiology. It's why ancient skeletons hold so many secrets about the lives of our distant ancestors, from the skeletal changes we see in the skeletons of the longbowman of the Middle Ages to structural changes to the pelvis and spine. These and many

**Opposite page:** Coloured transmission electron micrograph of cardiac muscle fibrils from a healthy heart. Mitochondria (yellow) supply the muscle cells with energy. The muscle fibrils, or myofibrils, are crossed by transverse tubules. These tubules mark the division of the muscle into contractile units, or sarcomeres.

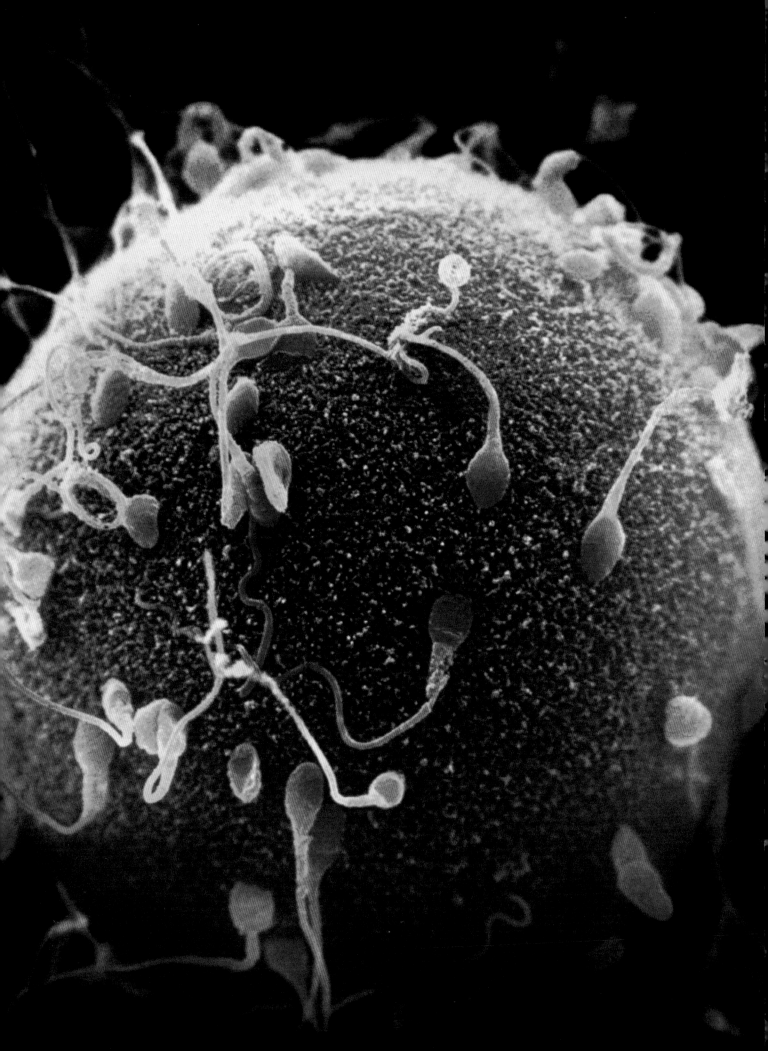

other examples reveal a growing trend: as we look to unlock more and more of the secrets of the human body, we are finding that the solid lines between nature and nurture are rapidly beginning to blur and nowhere is this more apparent than in the field of genetics.

# NATURE NURTURED

Each one of us has a day some 38 weeks or so before our birth-day that marks our beginning. It's a day that for many of us is lost in the history of our parents' lives at a time before they even knew we existed. The result of a series of events that occurred on average 268 days before our birth. That day is of course the day of your conception, the moment sperm met egg and the 23 chromosomes from your father combined with the 23 chromosomes from your mother to create a new human – you.

For over 150 years, since the Czech monk Gregor Mendel first began revealing the laws of genetics with his garden of peas, we have come to realise the power of that genetic fusion. At that one moment a multitude of traits, be they physical, physiological, psychological or pathological, are set in stone, a blueprint for your body and much of the life it will lead. At that instant much of your appearance is established, from height to eye colour, skin tone to the straightness of your hair. Your blood group and pulse are in some way defined at this moment. Your tendency to seek thrills, suffer from anxiety or even the age at which you are likely to lose your virginity are also all influenced by the assortment of genes you are dealt that day. The length and quality of your lifespan is laid out with a first preliminary draft of the illnesses you will endure through your life, from serious inherited conditions like cystic fibrosis and Huntington's, both absolute in their genetic determinacy, to the odds you will suffer from heart disease or certain forms of cancer. And that's not even the start of it – even your tendency to baldness if you are a man are all decided on that day. That's a hell of a lot that was decided on a day you might know nothing about!

The idea that all of this and so much more is set in stone at the moment of conception goes back to the genetic laws that Mendel first set out in his experiments in the middle of the nineteenth century. Although initially forgotten or ignored for almost 40 years after his discoveries, Mendel's work would ultimately

**Opposite page:** Coloured scanning electron micrograph of sperm on the surface of a human egg during fertilisation. Each sperm has a head, neck and a long tail. Here the sperm are attempting to penetrate the thick surface of the egg. The human female usually produces a single large egg, yet only one of the millions of male sperm released may penetrate the egg to fuse with the egg nucleus. Once fertilised, the egg begins its process of growth into a human embryo.

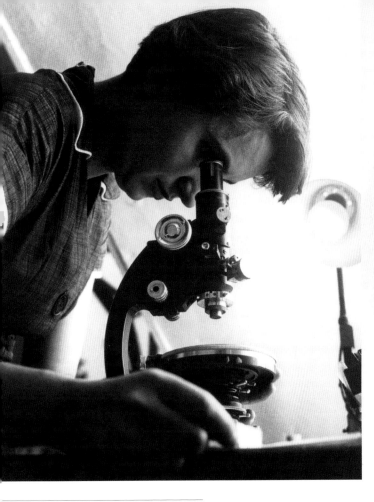

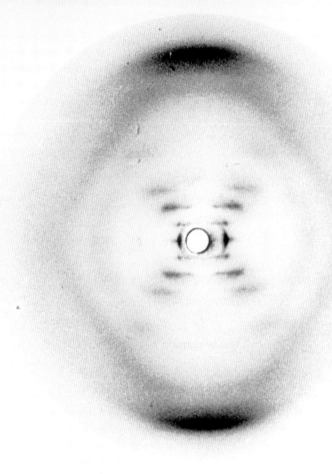

**Above left:** Portrait of Rosalind Franklin, British X-ray crystallographer, whose work producing X-ray images of DNA was crucial in the discovery of the structure of DNA. Her X-ray images of DNA crystals were of the highest quality and were crucial to Watson and Crick finalising the structure of DNA.

**Above right:** X-ray diffraction photograph of DNA obtained in May 1952 by King's College London researchers Franklin and Raymond Gosling. This image, commonly called 'Photo 51', is of the beta of DNA. The image results from a beam of X-rays being scattered by a crystal of DNA. Various features about the structure of the DNA can be determined from the pattern of spots and bands. The cross of bands indicates the helical nature of DNA.

set the foundation of our understanding of genetics through a description of trait-carrying alleles, which can be dominant or recessive and that are carried by the gametes of an organism, in our case the sperm and egg. Over the last 150 years we have built on these principles and added an astonishing array of knowledge to allow us to understand the mechanisms of inheritance in extraordinary detail. By 1902 the chromosome had been identified as the carrier of genetic material through the independent work of two men, Walter Sutton and Theodor Boveri. Their studies on sea urchins and grasshoppers paved the way for Thomas Hunt Morgan to provide irrefutable evidence for the role of chromosomes in heredity using the drosophila, or fruit fly, as his experimental model. By 1915 Morgan's fruit flies had cemented the idea that chromosomes are the carriers of inheritance and that they contain individual genes located along specific sites of the chromosome that act as the basis of heredity. It would, however, be another 38 years before the secret structure of a chromosome and the genes it contains would finally be revealed. A two-page paper published in the journal *Nature* on 25 April 1953 by Francis Crick and James Watson would share the secrets not just of human inheritance but the inheritance of all life on earth through the description of a molecule called Deoxyribose Nucleic Acid, otherwise known as DNA. Their work, together with the work of Rosalind Franklin and Maurice Wilkins, revealed the processes that underlie the genetic inheritance of every living thing through a code using just four different components, the nucleotide subunits known as bases – adenosine, thymine, cytosine and guanine. This simple code written in four letters, A, T, C and G, contains all the instructions to turn a single cell at the moment of conception into the trillion cells that make up a new-born human being. The ultimate secret of the human body seemed to have been revealed in this discovery, a door opened to the library containing the blueprint for a human life. It would take another 50 years of innovation and investigation before we believed we could read this blueprint in full but by April 2003, with the completion and publication of the first draft of the Human Genome it really did seem as if we had access to every single secret the human body contained. But since we reached that apparent milestone, it has become clear that even with the complete human genome at our fingertips, the process by which inheritance functions is far more complex and far more entwined with our lives than we imagined.

Over the last decade or so, geneticists have begun to understand that the moment of conception is far less the locking down of a genetic blueprint for your life and far more a genetic work in progress. This emerging field of epigenetics is revealing that your DNA itself is not a static, predetermined program but instead can be modified and controlled by your environment and the biological markers that are switched on and off by the specifics of the life that you lead. In other words, for good or ill, your genes are adapting, changing and learning from the environment that you place them in. Your genes, your inheritance, your nature is very much being nurtured.

# LIFE IN THIN AIR

Breckenridge, Colorado, is home to the highest ski resort in North America. Almost 4,000 metres above sea level the peaks of the Tenmile Range, part of the northern Rocky Mountains, provide a spectacular backdrop for those who come in search of its famous slopes, but at this altitude the effects can be significantly difficult for many of those who arrive here. The town of Breckenridge itself sits a little further down the valley at almost 3,000 metres but even at this altitude the effect on the human body can be quite profound. The air in Breckenridge is only three quarters of the density of the air at sea level and thus in any given breath there is a 25 per cent reduction in the number of oxygen molecules present. Altitude sickness is an unpredictable business. Some of us can begin to feel its effects at elevations as low as 1,500 metres but acute mountain sickness, or AMS, most commonly occurs above 2,500 metres. This means that around 20 per cent of the people visiting Breckenridge will suffer from a variety of symptoms that can include headaches, fatigue, swelling of the hands and face and insomnia.

This is exactly what happened to Mark Benson when he moved here from the lowlands of Ohio in 2015. Setting up a construction business in a new town is not easy when you're feeling the effects of AMS and, combined with the fact that Mark was also hoping to become part of the local mountain rescue team, it was not an ideal start to a new life. However, within just a few weeks all of the symptoms of AMS had disappeared and Mark found that his body had completely acclimatised to life at altitude. Unaffected by the lack of oxygen, his body has effectively adapted and learnt to live with the oxygen that is available.

It's long been known that humans can adapt to life at altitude. Throughout the twentieth century, high-altitude populations such as the Quechua people of Peru and the Sherpa communities of Nepal have been intensively studied in an attempt to understand their specific physiological adaptations. The evidence now strongly suggests that at least for the Sherpa community, the adaptation is a rare observable example of natural selection playing out in humans. These communities have inhabited the high Himalayan plateaus for over 3,000 years and during that time it seems respiratory traits that benefit life at this altitude such as lung volume, respiratory rate and cerebral blood flow have been selected for in the population. Over thousands of years the pressure of this extreme environment has turned from nurture into nature – Tibetans are now born ready for life in thin air.

This kind of inherited adaptation takes thousands of years, but for the rest of us lowland dwellers, the human body has an amazing ability to deal with the challenges of high-rise living. Remarkably the very latest scientific research is suggesting that even in the short term our genes are able to adapt and learn from the environment, an example of genetic flexibility that just a few years ago would have been unthinkable.

For someone like Mark arriving in Breckenridge from Ohio, in the short term, the journey from low to high altitude is dealt with by an immediate physiological response to keep him alive. A reduction in the level of oxygen in his blood triggers the body to maintain its homeostatic balance by increasing the depth and rate of breathing, increasing his heart rate and shunting blood away from areas like the gut and towards the heart and brain. This increased breathing removes acidic carbon dioxide from the blood, making the body slightly alkaline. This means his kidneys have to compensate by removing alkaline bicarbonate from his blood. Slowly over the following days and weeks, his body begins to mount a more long-term response to the new challenges it is facing. On average it takes approximately 11.4 days' acclimatisation for every 1,000 metres of altitude, so for Mark's body to acclimatise to life at 3,000 metres in Breckenridge, it will take around five weeks to complete.

Two girls from Uros row a reed boat to one of the unique floating islands of Lake Titicaca, Peru – often called the 'highest navigable lake' in the world, with a surface elevation of 3,812 metres (12,507 ft).

During this time his physiology will markedly change. Inside his body the concentration of capillaries in his muscle tissue will increase to boost the efficiency of oxygen delivery, the right ventricle of his heart will grow larger to send greater volumes of blood to his lungs, which in turn will modify the flow of blood around the lungs to maximise oxygen transfer. All of this will happen over those first few weeks as well as metabolic adjustments, all to help manage the efficient uptake and use of the available oxygen. There is also another pronounced change to the physiology that is crucial for adaptation and one that elite athletes have long used to increase performance by training at altitude.

Gradually over the first few days and weeks, the lower levels of oxygen will stimulate his kidneys to produce higher levels of erythropoietin (EPO), a glyco-protein hormone that controls red blood cell production by stimulating the bone marrow to produce erythrocytes. This increase in circulating red blood cells raises the oxygen-carrying capacity of the blood, meaning Mark's muscles to get all the oxygen they need. The physiological response that produces higher levels of EPO involves a change in the genes that are controlling the production of EPO.

Robert Roach is the director of the Altitude Research Center at the University of Colorado Medical School, one of the world's leading centres for altitude research. Over the last decade Roach and his team have led the way in exploring the human body's response to the hypoxic (low oxygen) effect of altitude. In a series of innovative experiments, including sending a group of elite athletes to the top of Bolivia's 5,421 metre high Mount Chacaltaya, they have revealed that many of our assumptions about the body's response to hypoxia have been incorrect. For instance, the Chacaltaya experiment showed that altitude adaptation can be far quicker than previously thought, with changes to the ability of red blood cells to cling on to oxygen occurring within hours rather than days. Intriguingly this study also showed that the body 'remembered' these changes and so even when the volunteers left the mountain for one or two weeks, when they went back up they acclimatised much quicker, as if they had learnt how to respond. Understanding the mechanism for such rapid metabolic adaptation has led Roach and his team to the emerging field of epigenetics.

Epigenetics refers to changes in gene function that are not explained by changes in the DNA sequence. Instead they are changes to a layer 'above' the genes that act like switches, increasing and decreasing gene activity. The intriguing thing about epige-netic change is that unlike changes to your DNA sequence that require the long, slow process of natural selection to alter it, epigenetic factors can be changed quickly and directly by the environment. We have understood that different genes are switched on and off in different tissues for decades. This controlled by a fairly well understood

**Opposite page:** Xand climbing with the Xtreme Everest research group on Cho Oyu, the sixth highest mountain in the world, shortly before turning back with altitude sickness.

process of 'transcription control'. Proteins called transcription factors bind to regions of the DNA and recruit the machinery that produces mRNA that can then be turned into a protein. Epigenetic changes add a further, much less well understood layer of control. But, remarkably, there is some evidence that these changes may be heritable. With an epigenetic layer of control across our genes that is reacting to the environment, and in some cases may be passed on to future generations, we begin to see that *how* we live has a more direct impact on how our genes are expressed.

In the case of altitude adaptation Roach and his team now believe what we are seeing in the adaptive process is that the genes are learning from the new environment and undergoing an epigenetic change. Chemical groups are added to the string of DNA which allow it to be more easily switched on or off. When someone like Mark makes his new home in Breckenridge the research now suggests that this results in an *epigenetic* change to the DNA that codes for EPO. Inside his kidney cells, over a hundred methyl markers have been added to the genes that regulate EPO production. These methyl markers are tiny carbon-hydrogen instruction packs that bind to a gene and say 'ignore this bit' or 'exaggerate this part'.

Anecdotally this may explain the unusual age demographic of extreme high-altitude mountaineering without oxygen. There are a few dozen climbers in the world who have climbed multiple peaks above 8,000 metres – there are only 14 such mountains in the world and Mt Everest, whilst the tallest, is considered one of the easier ones. A particularly elite club is people who have climbed all 14. Whilst many of these climbers start in their mid-thirties, the hardest peaks are often climbed in their fifties or beyond. This may be due to epigenetic modifications that make ascent to high altitude easier with each climb. In my own small climbing career I found that as a teenager I got sick in the Alps. In my mid-twenties I was climbing over 6,000 metres in the Himalayas and by my thirties I was able to climb over 7,000 metres – although I spent a night unconscious in a tent with Xand and two other medical colleagues before deciding that whatever epigenetic modifications I had made, it would be unwise to try to go any higher.

These markers change the way the genes are read – so that EPO production in Mark's body is increased long term. This effect happens after five days at over 7,300 feet altitude and new science shows we have an epigenetic memory – methylation persists for five days or more. Roach describes this as a '*memory*' for high altitude. When the subjects of his studies went back to the high altitudes over and over again, each time their adaptation was easier and easier. 'We see hints of that in our data when people go down after acclimatising, then come back up,' Roach explains, 'there seems to be protection but also a re-ignition of the gene expression program ... so you're doing what you need to do faster.'

Although still in its infancy, the emerging studies into epigenetic effects provide a tantalising new insight into the complex balance between nature and nurture. We are now understanding, however, that environment shapes us in ways that are more complex than we previously imagined. It seems our genes are being modified by myriad environmental factors – stress, trauma, exercise and obesity, for example. Clues are starting to emerge that they could even be passed down through two or even three generations. Epigenetic inheritance is far from proven – but scientists are starting to get some tantalising results. Most evidence comes from work with mice and rats – in one study, DNA methylation caused by stress led to three generations of obesity and similar responses have been seen in nematodes, pigs and fish.

Your genes are learning, adapting and changing to the stresses and strains of your life and these changes can last days, months or even a lifetime. It means that the life your grandfather or grandmother led may well be affecting your health and well-being today, and equally the life choices you have made may well be felt by your own children or grandchildren. It remains to be seen how deeply these epigenetic changes affect future generations and us, but there is no doubt we have revealed a new and surprising secret link between our genes and our lives.

Chris ice climbing in Greenland. Virtually at sea-level, in fact.

# LEARNING ... TO FIT IN

From birth we learn at an extraordinary rate, and this enables our brains to perform more complex skills than any other animal on the planet, from walking to talking, to equipping us with the memories upon which our identity, our culture and ultimately our civilisations have been built. Our bodies are endlessly learning as well, carrying with each one of us a record of the lives we live, both in the very structure of our bodies and in our genes, as well. The environment influences every aspect of human biology, but there's one type of learning that we all share – we learn to exist, not just as individuals but in groups, tribes and communities.

We live in the largest communities of any vertebrate on the planet. From small villages to megacities like Tokyo, Los Angeles or Bejing that hold together tens of millions of people, living lives of extraordinary coordination and cooperation with each other. The remarkable thing is not that there are so many rulebreakers, but that there are so few. To make these communities function requires an ability to navigate a complex social world: a skill we must all learn. We learn to read the emotions and thoughts of others and in return we expect others to read and understand the emotions we express.

As a human yourself, you are almost certainly an expert at the reading of human emotions: a highly trained, sophisticated observer of the smallest expressions. Standard psychology used to suggest that human emotions can be broken down into six basic facial expressions. When subjects were presented with photographs of these and asked to read the faces, many very quickly labelled them as 1) Disgust, 2) Fear, 3) Joy, 4) Surprise, 5) Sadness and 6) Anger. Modern-day psychology no longer relies on just these six expressions as the basis of emotional expression. Paul Ekman, the eminent psychologist from UCSF, has shown that each one of us has the ability to communicate with over 10,000 unique facial expressions, with at least 3,000 of these communicating emotion. What's extraordinary about the complexity of non-verbal communication is how universal it is with diverse cultures across the planet recognising and agreeing on the nuances of emotions being expressed by a face. The universal language of human emotion is expressed most clearly in two  of the most powerful expressions of human emotion – laughter and tears.

Crying is a social skill we learn from a very early age, but its power remains potent throughout our lives.

A wild mountain gorilla in the Bwindi Impenetrable Forest, in Uganda. Our close hominid relatives communicate richly through a combination of facial expression and gutteral vocalisation.

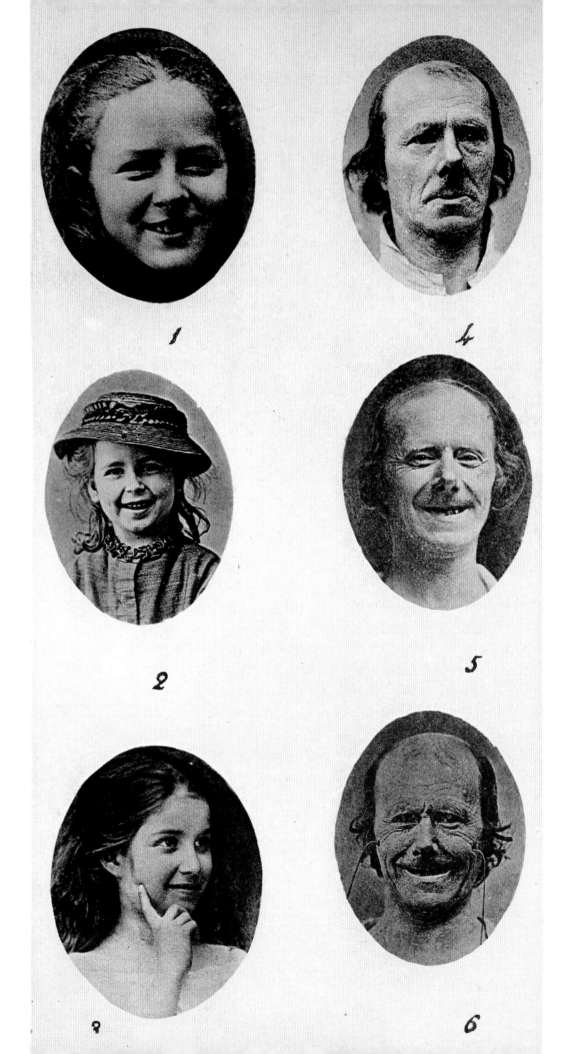

1

2

8

4

5

6

My daughter's wailing didn't have any real power until week 5, when she became able to produce tears – and get me and her mother to do whatever she wants. Exactly why you produce tears when you cry is unclear. In 1982, Dr William Frey measured cortisol levels in emotional tears and hypothesised that the body was using tears to eliminate the stress hormone. Perhaps this is why we feel better after crying. On the other hand, most tears flow through the lacrimal duct into the nose and are swallowed, so this explanation needs some tidying up.

Laughter is an equally important social tool. We all laugh, no matter where on the globe we come from. Laughter, real or fake, is a 'social emotion' that brings us together and helps us to bond, whether or not something is actually funny. That's because when you laugh with people, you are showing them that you like them, you agree with them, or that you are in the same group as them. In fact you are 30 times more likely to laugh if you're with somebody else than if you're alone – and we catch laughs most readily from people we know, and when we do catch a laugh, our whole body responds.

Heart rate and pulse elevate and then laughter begins to hijack the body parts that we use for breathing and talking. Rhythmical, often audible contractions of your diaphragm and other parts of your respiratory system make you go 'haha' or 'hoho'. Respiration rate and ventilation becomes irregular as an effect of the epiglottis half closing the larynx. The normal cyclic breathing pattern is disrupted: laughter is, in fact, in direct conflict with breathing – when people protest in hysteria, 'Stop, you're killing me!', they're not so far from the truth.

Laughing is so important to us socially that we have developed a special kind of laughter to use in group situations, This 'posed laughter' is voluntary and strategic, less authentic tones are often more nasal, but whether it's posed or real, laughter makes us feel good. It's often said that laughter is the best medicine and slapstick humour can be a painkiller – a good laugh makes the body release painkilling chemicals called endorphins.

Trying to understand emotions and emotional behaviour, like laughter and crying, begs questions about the nature of consciousness itself. Why do we need to feel any emotions at all? But there is no doubt that any behaviour so universally present must have provided an evolutionary benefit to our ancestors – allowing them to function in our complex social world.

**Opposite page:** 'Smiling' heliotype illustration of expressions from *The Expression of the Emotions in Man and Animals* by Charles Darwin, 1872. This was one of the first books to be illustrated with photographs with seven heliotype plates. The publisher John Murray warned that photographs 'would poke a terrible hole in the profits' and so the heliotype was chosen as a photomechanical process that could reproduce photographs directly on to the page (without mounting actual photographs). Darwin wanted to show the animal origins of human expression and emotion. He saw photography as a tool for accurately capturing brief expression.

We are not amused. Queen Elizabeth II cracks up US president Ronald Reagan.

# LIKE TO BE LIKED

Whether it's our first day of school or an awkward work event, our urge to fit in can seem all-consuming, but why is belonging to a group so important to us? A recent study by scientists at UCLA found that fitting in is neurologically similar to eating chocolate.

In a unique social media experiment, the UCLA team scanned the brains of 32 teenagers while they were browsing 148 photos, 40 of which they had submitted themselves. Each photo displayed the number of likes it had received, although in reality this number was being manipulated by the researchers. Unsurprisingly,

---

WHEN PEOPLE PROTEST IN HYSTERIA, 'STOP,
YOU'RE KILLING ME!', THEY'RE NOT SO FAR FROM
THE TRUTH ...

---

when the participants saw one of their photos with lots of likes, they found that the reward centres of the brain, in particular an area called the nucleus accumbens, were highly activated.

This finding might not be that surprising on its own, but the interesting thing about social media is that it gives us a quantifiable way of measuring how much being liked influences a group. And it seems the buzz we get from social acceptance can affect us in ways that are out of our control. When the participants were asked to 'like' any of the photos, they were hugely influenced by the number of likes the photo had. The same photos were shown to half the teens with lots of likes and half with very few. This meant that participants would see images either with lots of social endorsement, or barely any at all. Interestingly the people who viewed an image with many more likes were much more prone to like it themselves – even if this was by someone they had never even met, and even though the majority of participants claimed not to have noticed or been affected by the number of likes.

Our decisions are actually being shaped by our peers without us even realising. and it seems this effect is strongest in adolescents. This so-called peer influence is much more marked in adolescents – it's a critical time in our lives when the frontal parts of our brains involved in our social decision-making and reasoning are still maturing and developing. We need to learn to interact in the world around us.

So our innate urge to fit in may well make us open to influence.

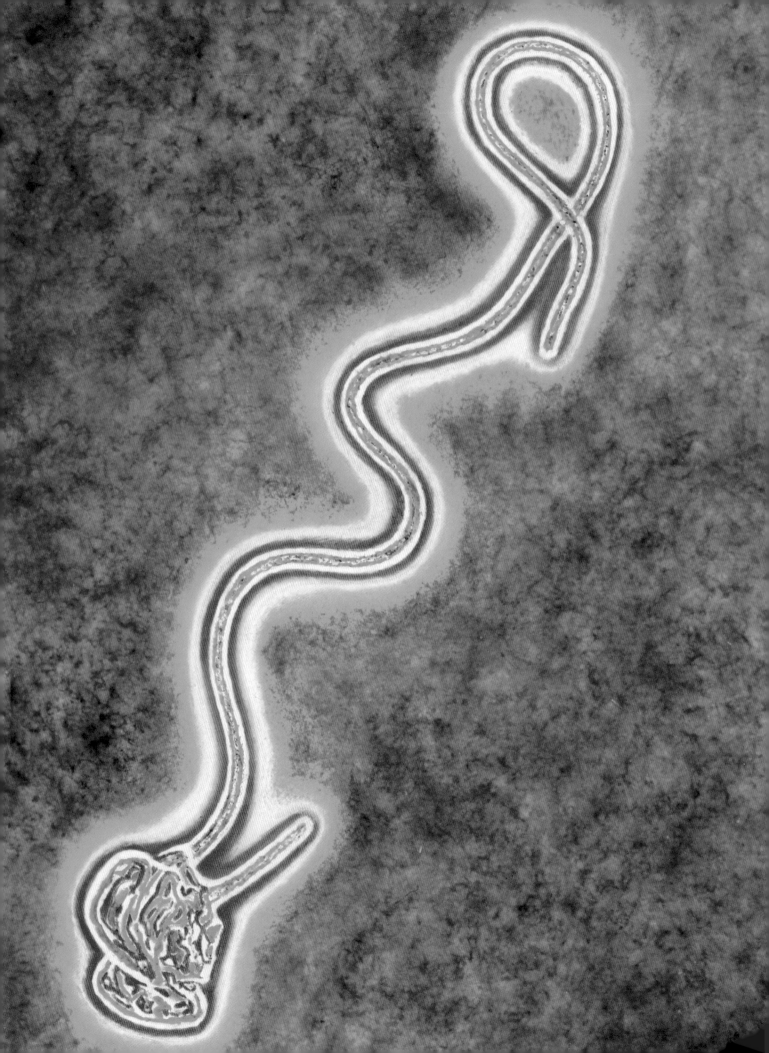

# THREE

**SURVIVE**

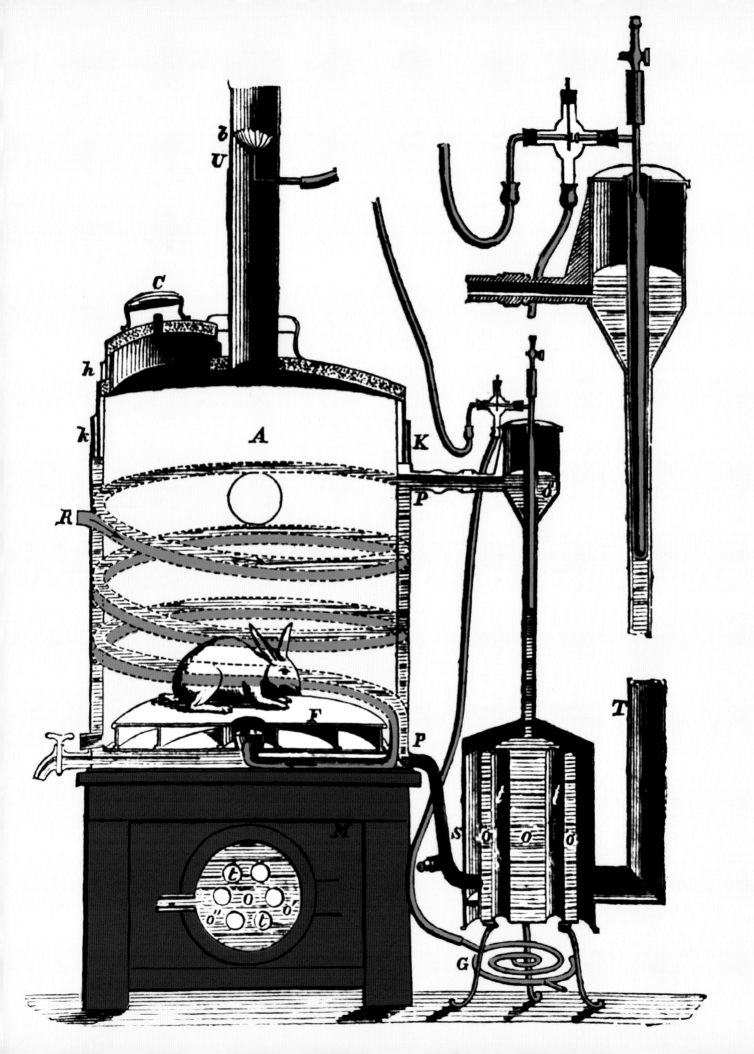

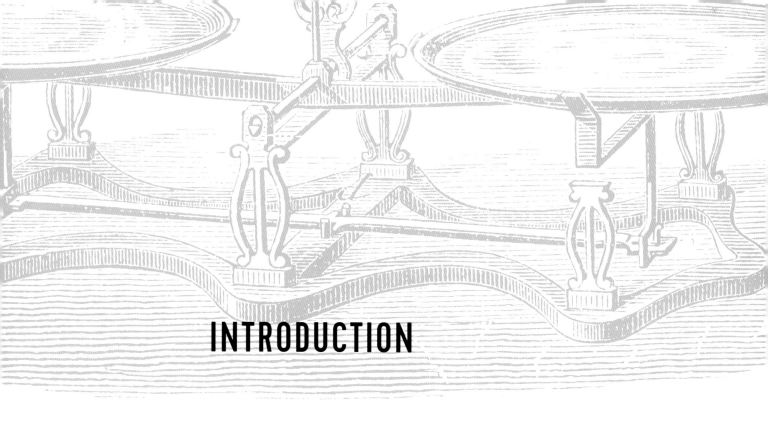

# INTRODUCTION

The chances are you're reading this in a room at 18–22°C, you've recently had a meal and there are no large predators nearby. But don't be fooled. It's a dangerous world out there. From your first breath to your last, your body is in continuous battle to maintain the specific conditions required for life. This battle is called homeostasis – quite literally meaning 'staying the same'. It describes the tendency of the body to maintain its internal conditions even when faced with external changes.

Why is this such a vital foundation for life? The answer can be found by looking at the wide variety of critical parameters that exist inside our bodies, which if varied by the smallest of margins rapidly result in death. Temperature, pH, oxygen and glucose are a few of the most familiar ones, but there are many more, and they all only need to change slightly for the results to be catastrophic. Each one of our cells is a complex bag of chemistry with specific reaction conditions. Our bodies are locked in a continual dance with our environment, which provides a relentless challenge. Twenty-four hours a day, 365 days a year our bodies monitor, react and adjust our relationship with the external world to make sure the internal world stays the same. This drive for sameness covers our heartbeat, blood pressure, urine

**Opposite page:** Apparatus developed by French physiologist Claude Bernard (1813–78), to study the heat balance of individual animals. *Milieu intérieur* (later dubbed homeostasis) is the key concept with which Bernard is associated.

output, calorie expenditure and many other processes devoted to homeostasis.

But homeostasis is not merely physical. It is intimately wired to our emotions. Fear is a system designed to alert and trigger evasive action whenever our internal balance is threatened, whether faced with the toxic threat of a poisonous snake or the gravitational threat of a cliff edge. Disgust is another homeostatic emotion, to prevent infection and infestation from the food we eat, contamination that once inside us can rapidly destabilise the balance. And when these early warning systems are breached and pathogens gain entry through the food we eat or the air we breathe, our immune system is equipped to react to anything that threatens the balance.

From the moment of conception, every cell needs stability. And this is when the battle for survival begins.

# IN THE BEGINNING ...

I have been present at one conception but I didn't witness it. The only conception I have ever witnessed was in the basement of a large Georgian house on London's Harley Street (or the street of shame, as those of us without an office there like to refer to it). Like almost every building on Harley Street, this was a private medical practice, in this case specialising in fertility.

In the basement laboratory a fertility technician produced a 10 cm petri dish. There were two drops of liquid in the middle of the dish – one a drop of semen, one a drop of fluid containing a few eggs. Almost all human sperm and eggs throughout history have been discarded as waste through menstruation or masturbation. The sperm in the dish we were looking at, even to the inexpert eye, didn't inspire much awe. Misshapen, barely moving and sparse, only a few were visible, twitching under the microscope. The egg, vast by comparison, looked equally unremarkable. Much less interesting than the cells I take care of in similar dishes in my own lab – cells vital to my experiments, cells which I have genetically manipulated and cared for.

The sperm was, obviously, about to be injected into the egg. I wasn't bored but I can't claim that I was fascinated either – I'm not generally moved by a sense of occasion. I didn't cry at my own wedding. Or graduation. And I felt like I'd seen this before. It's an image that's so often used it's become a visual, scientific cliché.

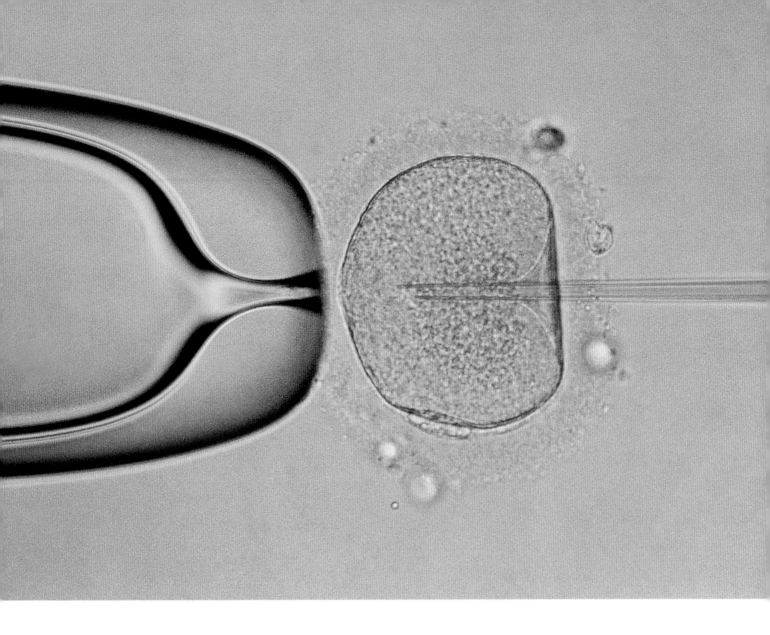

Matt, the director, wanted me to say something to camera about the 'spark of life' at the moment the sperm was injected, but I wasn't feeling it. Until the technician started to manipulate the sperm. On either side of the microscope slide was a system of levers connected to two hollow glass needles little thicker than a human hair. Under the microscope these looked like huge glass pipes. The one on the left held the egg, gently sucking it into the end of the needle like a tennis ball on the end of a vacuum cleaner. Next, with the right one, she used the needle to cut the tail off the sperm to 'activate' it, then sucked it up and stabbed the needle into the egg. This required some effort, the egg membrane distorting under the pressure of the needle tip before it was penetrated. Then the sperm was injected and ... there it was! The spark of life. The dish transformed from carrying waste products to carrying something living. A human cell with a sex and the full complement of genes to build a body that would last 80 years. Probably more. A baby conceived

**Above:** I've been present at two conceptions but only actually witnessed one. This is the moment when the sperm is injected into the egg.

**Opposite page:** Cryogenic preservation of frozen sperm straws and embryos in liquid nitrogen.

on Harley Street today must surely have a life expectancy of over 100.

But more miraculous to me was the conditions that this cell would need to be kept in for those 100 years. It was, at the moment, in the dish, teetering on the brink of death. The lab was tightly controlling a small set of variables the disruption of any one of which would be rapidly fatal. The acid, oxygen, carbon dioxide and salt concentrations in the dish cannot vary by more than a few per cent. The room was warm, around 28°C , and this is as cold as that cell will be for the rest of its life. Once implanted in the mother, the cells that this fused egg and sperm will divide to produce will maintain almost these exact conditions for decades. A few cells deep in muscles will tolerate greater acidity, temperature and altered salt concentrations for short periods during exercise. Cells very near the surface of the body will vary in temperature by a few degrees. But for the most part these variables will remain unperturbed regardless of the journey of the human they are in. This embryo may climb Everest, cross the Sahara or go to Mars – but to a cell in the middle of the body these extremes will be undetectable.

Xand goes for a run in Professor Mike Tipton's world-class extreme environment lab.

# SOME LIKE IT HOT (OR COLD)

This ability of the body to defend its inner conditions is the focus of interest for a very different lab at the University of Portsmouth run by physiology professor Mike Tipton. Mike is an expert at disturbing the tranquillity of the body's core in his world-class extreme environment lab. In a single room here he can have you run a marathon that starts at the North Pole in winter and ends in Death Valley in summer. If you're still game (or if he's in the mood), you can then get in a flume pool filled with ice and simulate swimming the Bering Strait. Any protestation that you've had enough is quickly dealt with according to the data.

A claim that you're 'too hot', for example, can be disproved instantaneously by the readout from the thermometer capsule you've swallowed. Mike will cheerfully inform you that you're still a few degrees away from death so can safely continue, adding earnestly that of course you can stop at any time, for any reason. Equally, claims that you pushed yourself to total exhaustion can be verified by looking at your oxygen consumption. Most of us, it turns out, are nowhere near our body's limits when our minds fail us. I've been to Mike's lab a lot over the years and he's become a friend, although seeing him is always horrible.

But not today. Today is a happy day for me because I have brought Xand so I will be playing the part of control while Xand goes for a run. He's going to start at -20°C and finish at over 30°C. It's worth adding that Mike takes safety and ethics very seriously – informed consent is given and we are in good hands.

Mike wants to show us the power of homeostasis to protect the body from extreme environments and also how your body protects you from yourself. This is a particular problem with temperature regulation, since the human body can only operate within a very, very tight temperature range and it generates immense quantities of heat. Mike explains:

> When you're a small organism your main problem is external influences, but as you get larger the problems of homeostasis become internal not external. The human body is essentially a bag of primordial soup. Salty water with chemical reactions going on. As the bag gets bigger it gets harder and harder to get rid of acidic waste and heat as fast as they're generated. For most of our evolutionary history we have lived in environments where the air temperature was almost body temperature, so the body has evolved as a superb heat loss machine.

It's also fairly superb at staying warm in extreme cold, but from Mike's perspective a jog along a beach on holiday is, theoretically, more of a threat to life than running in the Arctic. Even at rest, your body is generating enough wattage to run a bright old-fashioned light bulb, around 60–100 watts. If you get up to a sprint and you're extremely fit, you'll have the equivalent of a two-bar heater (or a powerful kettle) in your belly. In other words you'll generate enough energy to boil yourself unless you get very good at shedding it. In the Arctic it's pretty easy to lose heat. On a beach holiday, or in this case Mike's lab, at a balmy 30°C, it's much harder.

Today's jog will start in the high Arctic. The room has been pre-chilled to -20°C. Xand gets on the treadmill and starts jogging. He has swallowed a thermometer capsule that is sending his temperature by radio signal to a digital display in the lab. It is 37.5°C. His pulse is 110. We are watching him with a thermal imaging camera that can tell us the exact temperature of his skin.

Xand is wearing a hat and gloves but is otherwise dressed for the beach – shorts and a T-shirt. You might expect that blood would be diverted away from his hands to his body's core but such is the amount of heat generated that even at -20°C he takes the gloves and hat off after a few minutes. His hands are warm to the touch. Shortly after that he starts sweating, but instead of the sweat running down his face it forms a beautiful filigree of ice on his forehead at his hairline. Even at this temperature his body is diverting blood to the skin to lose heat and he feels totally comfortable. Survival in the cold is limited therefore by fitness and endurance. Over several expeditions I have walked thousands of miles on polar ice caps in little more than a woolly jumper and chinos. If you can keep walking, then you won't die. Cold-water swimmers prove this point even more conclusively. Water will strip heat from the body far, far faster than air but by being supremely fit and conditioned, people can swim for many miles in water so cold it is barely liquid. In fact one of the senior scientists at the lab, Heather Lund, has just returned from the world ice-swimming championships. The team here are prepared to suffer every bit as much as their subjects.

After 15 minutes jogging Xand's body temperature has risen by one degree. His hands are warm and his T-shirt is soaked with sweat. The only parts of his body showing any evidence of cold are his ears, which are now at less than 10°C. Mike stops the treadmill and allows Xand a safe (if not generous) recovery period at room temperature before moving him, dressed in the same clothes, into the hot room.

For reasons determined by physics rather than biology, if the air temperature is less than body temperature (37.5°C), you can cool off simply by having air blow over your skin. In theory at least. In fact you'd need a howling gale to lose heat

**Opposite page:** Thermoneutral: my huskies in Greenland sleeping peacefully through a katabatic storm. The ambient temperature was -30°C, with windspeed around 60 knots.

Chris swimming next to an iceberg in the Arctic Ocean. My fear of the creatures of the deep far eclipsed the discomfort from cold. My abiding memory is of the exaltation I felt afterwards and how I wanted to get back in.

in warm air, so you start sweating long before you get to body temperature even at rest. Xand jogging at an ambient temperature of 30°C has turned red and is starting to glisten with sweat.

But once the air temperature is over body temperature, it is impossible to lose heat unless you sweat. If you prevent the body evaporating sweat and thus losing heat, or if you generate heat quicker than you can shed it, then you die fast. A high fever is 40°C and very survivable. But very few people survive a core body temperature above 43°C.

Xand has been jogging now for around 10 minutes and is puce. His hair is matted with sweat and he is breathing much harder now to maintain speed. His pulse is now 130.

Normal body temperature is controlled astoundingly tightly. A naked man standing in still air will expend no energy regulating his temperature at around 25–30°C. This is our thermoneutral temperature. This temperature is comfortable, as it means skin temperature will be 33°C. This level of comfort is remarkably consistent around the world. You can ask a London stockbroker or a Kalahari bushman and they will all feel comfortable with a skin temperature of 33°C. And humans work very hard to achieve this with our behaviour, adding and removing layers of clothes and minutely adjusting thermostats and air-conditioning. Once the air temperature gets warmer than 24–28°C, you'll start to sweat. Any cooler and you'll start to shiver. This makes

███

'IF YOU ARE LUCKY ENOUGH TO SURVIVE LONG
ENOUGH TO DIE OF HYPOTHERMIA, YOU HAVE
DONE VERY WELL; MOST DIE IN THE FIRST MINUTE
OF IMMERSION.'

**PROFESSOR MIKE TIPTON**

███

us seriously tropical animals. A 1950s study took a series of Arctic and tropical mammals and birds from Alaska and Panama and heated them up and cooled them down in a chamber where heat production could be measured. They discovered that larger Arctic mammals (like an Arctic fox) didn't start to shiver until well below -30°C. My own experience of huskies on several Arctic expeditions is that they were totally content sleeping outdoors, naked (obviously) at -40°C.

If you put on some clothes, then a comfortable temperature to sit and work in is around 18–20°C. If the temperature of your environment increases, then the big change that allows you to lose heat is blood supply to the skin. Temperature sensors in the skin and muscles send information about core body temperature to the anterior hypothalamus, where the information is processed. The hypothalamus sends out signals that increase blood flow to the skin (as much as 8 litres/min) and increase sweating.

If you can't sweat, the core body temperature will start to rise dangerously. This is a particular risk for emergency services or military personnel wearing protective clothing. It doesn't matter how tough you are. If you work in an airtight suit (perhaps to protect you from fire or nuclear, chemical or biological warfare), you'll be able to work hard for about 20 minutes before you dangerously overheat.

Xand has now been jogging for some 20 minutes. He is gasping to maintain his speed. The armpit patches of sweat on his T-shirt have now met in the middle of his body. The treadmill is slick with sweat. His pulse is 160 now as his heart works not just to supply his muscles with oxygenated blood but also perfuse his skin with enough blood to lose heat in the still 30°C air of the lab. His core body temperature has risen by the same amount as it did at -20°C. A single degree. As Mike says, the body is a superb heat loss machine.

But Xand doesn't have long left. He can keep this up for a while but eventually he'll dehydrate. As this happens his sweat production will drop. His heart

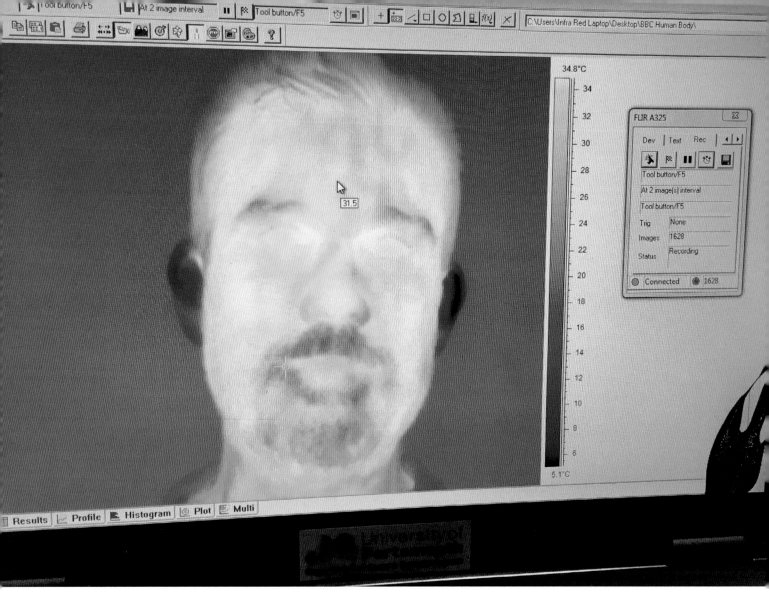

Xand at -20°C on Mike Tipton's
treadmill. Only his ears are cold.

will start to work harder to shift viscous blood to the skin. This increased effort will generate more heat. There will come a point where even if he stops exercising, he may have gone too far and his body temperature will keep increasing in a destructive spiral. Eventually at around 43°C he will start to cook himself. The heat will denature his proteins. The lipids that comprise the walls of his cells will turn runny and liquefy. His cardiovascular system will collapse, organs will fail and he will quickly die.

Mike explains all this, with one eyeball on the thermometer capsule reading, while Xand jogs along. The homeostatic effort is now evident as his pulse rises to 180 but as Xand hits stop on the treadmill his temperature is still fractionally lower than when he started. 'You could have gone on a lot longer ... in theory,' comments Mike, almost to himself.

As successive human migrations took populations from the equatorial and tropical regions of the globe to freezing mountainous and Arctic regions, we have made increasingly sophisticated behavioural adaptations. The modern urban environment is constructed so that there is almost never a challenge to core body temperature. But within all of us are a set of reflexes and adaptations that equip us to survive over a vast range of temperatures, tightly defending that core body temperature so that a cell in the middle of the body will remain entirely unperturbed by running in shorts at the North Pole or across the Sahara.

# A COLD WAR ...

When we think of maintaining balance in the body, we often think of the systems in isolation. One set of responses to hot, one to cold, another to low glucose or low oxygen – each triggers a different response to deal with the situation, from shivering, to sweating, to panting. In fact the body's response is far more interlinked and integrated, a system that is deeply entwined in the internal biochemistry of our bodies.

Every day of our lives this biochemical matrix is subtly challenged on a minute-by-minute basis, from missing lunch, to walking up the stairs, from walking out into a cold winter's morning to climbing a mountain, every external change requires an internal response however subtle. But sometimes that external challenge is so overt we can see the body's battle for survival writ large.

One such example can be witnessed every year in the ice fields of Norway. During the frozen winter months it is something of a popular Nordic tradition to cut a large rectangular hole in a frozen lake and submerge oneself in the freezing cold water for a minute or so. The benefits of this are claimed to be both psychological and physiological, reducing stress, fatigue and improving mood and also reducing the symptoms of conditions such as asthma and arthritis. I've done quite a bit of cold-water swimming over the years, mainly in the UK, and I love it. I don't believe the benefits are overstated. I've swum in ice holes in Norway and in the Arctic Ocean, but only as I toughened up in my thirties. I don't know if I would have coped with cold swimming as a rite of passage in the way that one group from Norway does.

Each winter a group of children from Eiksmarka, Norway, make their way out across the frozen lakes surrounding the town. Stripping to just their swimwear and socks, they stand on the edge of an ice hole and prepare themselves for the most intense minute of their young lives, a minute that will test their internal chemistry to the very limit.

The stoical children of Eiksmarka, Norway, prepare to jump.

Watching as these children jump into the ice-cold water, you might think the most immediate threat is all about keeping warm, but maintaining a stable body temperature is not the first hurdle they face, although ultimately it will be a pretty major one. Surprisingly, submerging yourself in water that is pretty close to 0°C does not kill you quickly; in fact studies suggest that it takes around 30 minutes to reduce your body temperature to a potentially deadly hypothermic state. Instead it's actually a complex interaction of bodily responses that provides the most immediate danger. Sudden submergence in ice-cold water elicits a response known as cold shock, a survival mode that if allowed to run out of control can lead to death within minutes.

Cold shock begins with the sudden drop in skin temperature, which, within a split second, triggers the blood vessels of the skin to contract. This diverts the warm blood away from the limbs and extremities to the core of the body. The reason? First and foremost the body's priority is to protect the vital organs by keeping them warm at the critical 37.5°C. This immediately puts a greater strain on the heart muscle, with the workload required to pump blood around the body increasing due to the vasoconstriction and the cold blood returning from the extremities also putting a greater stress on the organ's electrical function. It's no problem for a fit and healthy young body to deal with, but it's for this reason that those with heart disease are advised to avoid cold-water swimming, as it can rapidly trigger the heart to go into a state of electrical chaos or fibrillation and even cardiac arrest.

The other immediate response to hitting the cold water is an uncontrollable gasp for air. This inhalation can be so violent and sudden that it can cause you to breathe in water rather than air and this, not hypothermia, is the most common reason for drowning in cold water. But that's just the beginning of your problems. In the cold of the water our immediate response is to thrash around and shiver, an inbuilt reaction that generates heat from the additional muscle activity, but as we do this our muscles produce increasing amounts of carbon dioxide as a by-product of the additional metabolic activity and this floods into the bloodstream, increasing the acidity of the blood. A rise in blood acidity equivalent to just a drop of lemon juice in a glass of water would be fatal, because it has an immediate effect on the biochemistry of every cell in the body, but our bodies have multiple defence systems that maintain our blood pH second by second.

One of these is hidden away inside your arteries and brainstem where there are clumps of specialist nerve cells called glomus cells whose job it is to monitor your blood pH by detecting not only the levels of oxygen but more importantly the levels of carbon dioxide in the blood. If they detect a small change in the level of carbon dioxide and consequently the blood pH, then it's their job to send a signal to the medulla oblongata – the part of the brainstem that controls breathing. This alerts the body to shift its breathing activity to counteract the rise in pH by stimulating the diaphragm to make you breathe deeper and faster – in other words to hyperventilate. It's an unconscious rapid reaction to counteract any imbalance in the body's pH.

This is why on entering cold water that first initial gasp for breath is followed by a period of rapid breathing to expel excess $CO_2$ and maintain blood pH within safe limits. This breathing response in cold water is the same as drives increased breathing with any form of exercise. The urge to breathe is not directly driven by a lack of oxygen but is actually controlled by elevated levels of carbon dioxide.

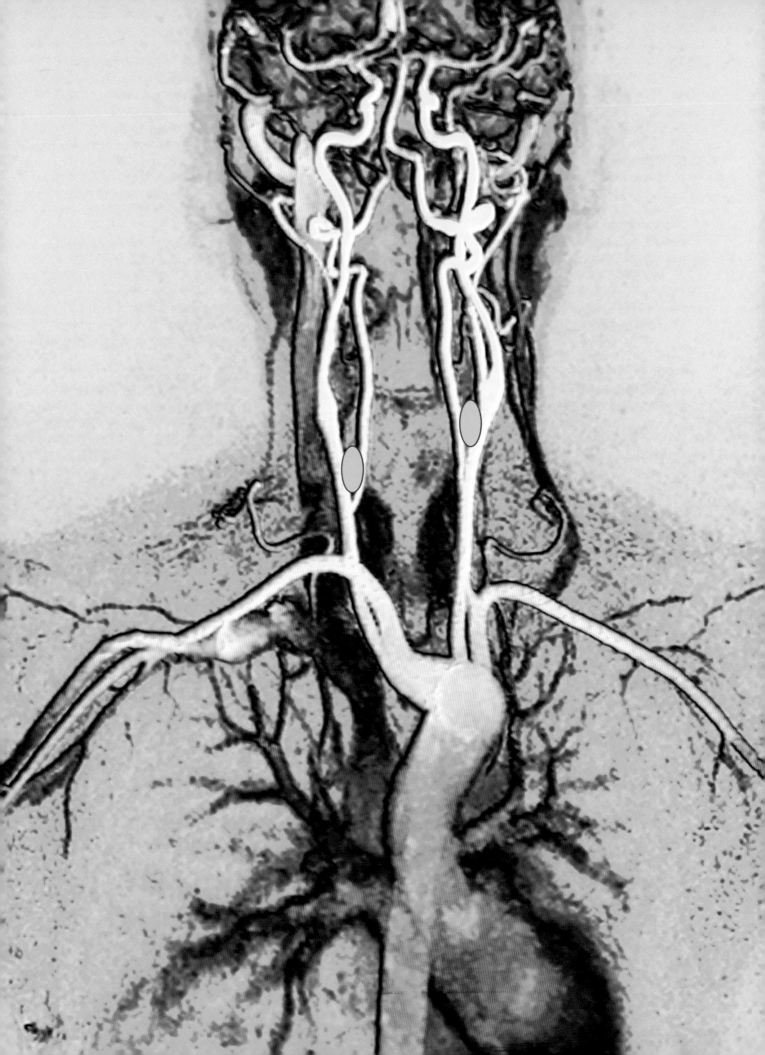

Of course, as the body responds to the increase in acidity by stimulating rapid breathing, it needs to be careful not to tip itself too far in the other direction by removing too much carbon dioxide, something that sounds beneficial but is actually dangerous. Reduced carbon dioxide in the blood is a condition known as hypocapnia, with the pH balance of the blood going from acid to alkaline. Just as a small upward change in the acidity of the blood can be deadly, so can a change in the other direction. And so, as with all homeostatic control, a balance has to be struck between the body's response and the ability to limit the effect, a process that is called a negative feedback loop.

Jumping into ice-cold water in the middle of winter is an extreme example of the body's homeostatic mechanisms, but it reveals the integrated systems that fight to keep your body in balance. The external shock may be driven by a sudden single environmental change, but the response requires an integrated reaction to coordinate a wide range of responses, all to keep the body in balance.

**Opposite page:** Coloured magnetic resonance angiography scan of a normal carotid system of a 45-year-old female. Bottom centre is the aortic arch, which curves over the heart. The arteries that branch off from these are: the brachiocephalic artery (left), the left common carotid artery (centre) and the left subclavian artery (right). The right and left common carotid arteries supply the neck and the right and left subclavian arteries supply the arms. The position of the carotid bodies are indicated by the green ovals.

# SCAREDY CAT

Maintaining this balance is so vital to our existence that not only is your body primed to respond to the smallest change, it also comes hard-wired with an early warning system whose primary purpose is to set off an alarm if we sense that the status quo may be challenged – an early warning system that we all experience as the emotion we call fear.

What are you scared of and why? Many of us have the odd specific thing that we consider scary – Xand and I share a deep and abiding fear of sharks. Xand used to fear spiders until he got a pet one called Doug. But beyond our own idiosyncrasies, many of our fears are remarkably universal. Ask a group of people in Britain or the US what their greatest fears are and you will not be surprised to hear that arachnophobia (spiders) and acrophobia (fear of heights) are commonly near the top of the list, with ophidiophobia (snakes) and aichmophobia (needles) not far behind. Despite the list appearing remarkably disparate, what connects these phobias is the fact that they all in different ways threaten our homeostatic balance. Whether it's the simple gravitational threat of falling from a height, to the risk of a venom unpicking our biochemistry, all of these threaten the finely balanced conditions within our bodies and need to be protected. To make sure of that protection,

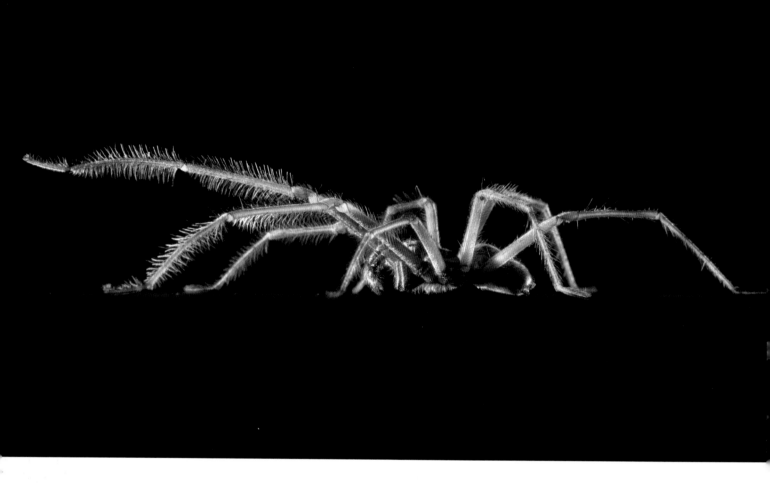

Male house spider (*Tegenaria domestica*) walking. Xand used to fear spiders until he got a pet one called Doug. Arachnophobia and acrophobia (fear of heights) are two of the most commonly held fears.

our reactions to our environment are consistently governed by the most primitive of responses – the empty feeling in your gut, sweaty palms, racing heart are all universal features of the response we call fear.

However, despite its importance for our survival, it seems we are not actually born fearing very much at all.

Thanks to a brilliantly odd study called 'The Visual Cliff experiment' conducted by Gibson and Walk at Cornell University in 1960, which involved encouraging crawling babies to make their way across a perceived drop, we've learnt that from the moment we can move we have a developed and seemingly innate fear of falling that is not just found in us but replicated across many species. The earliest expression of this is seen in a reflex known as the Moro reflex, a response to falling that results in a very specific set of limb movements and is thought to have evolved to help a newborn baby cling on to its mother when it momentarily loses its grip.

Apart from that there is only one other fear that we seem to carry with us from birth and that is a fear of loud noises. The startle reflex is a universal response to an unexpected sound that can also be seen from the very moment of birth. The reflex is triggered by sudden noises of around 80 decibels or more and results in a highly characteristic physical response that involves the stiffening of the arms and legs, hunching of the shoulders to protect the neck and the tightening of the body wall to brace for an incoming 'attack'. The innate nature of this fear of loud noise was most famously exploited in a series of notorious experiments known as the 'Little Albert' experiments conducted by the American psychologist John Watson, of Johns Hopkins University, in 1920. Legally and ethically the experiment would never be allowed to happen today but back then Watson gained permission to take 'Albert', a 9-month-old perfectly healthy baby boy, and explore whether the fear generated by the startle reflex could be associated with other stimuli. Just like Pavlov and his salivating dogs, Watson conditioned Albert to associate his fear of loud noises with the presence of a white rat. To start with, the child was completely unfazed by interactions with the animal but after being exposed to a loud distressing noise (created by striking a steel bar with a hammer behind his back) every

The Visual Cliff Experiment, developed by Eleanor Jack Gibson, showed that from the moment we can move we have a developed and seemingly innate fear of falling.

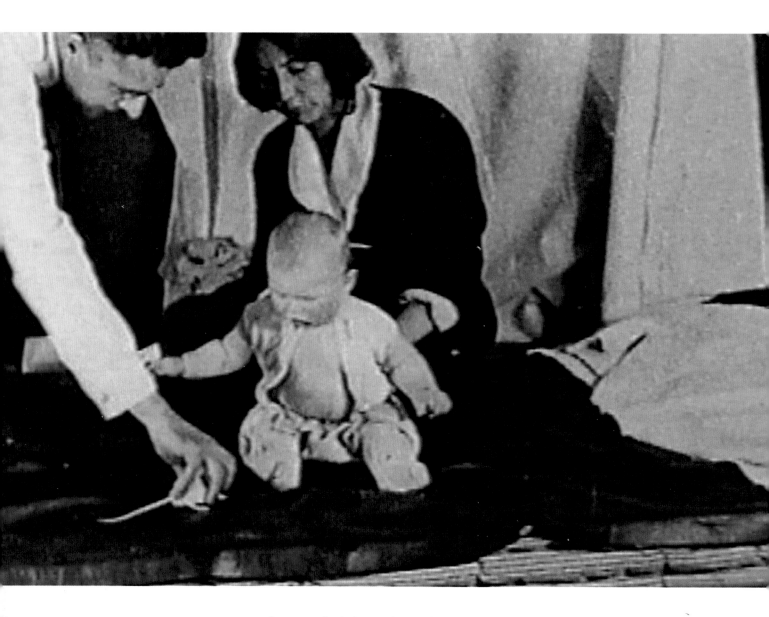

One fear that we seem to carry with us from birth is a fear of loud noises. The innate nature of this fear was most famously exploited in a series of notorious experiments known as the 'Little Albert' experiments conducted by the American psychologist John Watson, of Johns Hopkins University, in 1920.

time Albert touched the rat, he gradually developed a deep seated associative fear that resulted in the boy becoming distressed and anxious every time he saw the rat even in the absence of any sound.

These experiments, along with many other less controversial subsequent studies, have not only confirmed the fact that we are born pretty much fearless of most things but also given us an insight into how we slowly learn our set of fears from our experiences and from those people around us as we grow.

So what is fear? To understand the biological basis of fear, we need to look at some of the most ancient structures inside our brains, structures that are fundamental to our fear response and found not only in all mammals, but reptiles and birds as well.

# THE WOMAN WHO FEARED NOTHING ...

As we've already seen earlier in the book, our understanding of the human body has often been advanced by what's revealed when both the body and particularly the brain are damaged by accident or disease. When it comes to fear, one case in particular has opened our eyes to a deeper understanding of the way this emotion is generated in the brain.

SM, as she is known in the medical literature, is one of the most studied and cited individuals in medical history. To meet her you would be hard pushed to notice anything exceptional about her appearance or behaviour. She is a mother of three children, married and by all accounts living happily somewhere in the United States. Her identity remains a secret to protect her anonymity and she has never spoken publicly, but recently she agreed to an interview with one of the scientists who has worked with her most closely. In the interview, which was later released for broadcast to public radio, Dr Daniel Tranel of the University of Iowa explored the intriguing life of the woman who quite literally cannot feel fear. When asked to describe what fear is, she answered 'to be honest, I truly have no clue'.

SM was born with a rare condition called Urbach–Wiethe disease. This genetic disorder has only ever been diagnosed in a couple of hundred people since it was first described in 1929 and the symptoms vary hugely in each case, from abnormalities of the skin and mucous membranes (such as the lining of the mouth) to neurological abnormalities caused by the hardening of brain tissue. The genetic fault that defines the disease is thought to lead to the defective production of a key protein involved in the building blocks of cells. In the case of SM this fault expressed itself not in her body but in a very specific part of her brain. During her late childhood it transpires that the disease caused the discrete and complete destruction of an area of her brain known as the amygdala. This almond-shaped area of the

This coloured positron emission tomography (PET) scan of a transverse section through a human brain shows the response to fear. Subjects in this study viewed a series of photographs of a man's face depicting increasing stages of fearful expression. The active brain region colour coded yellow/red is the left amygdala, suggesting it is this part of the brain that recognises and responds to facial fear.

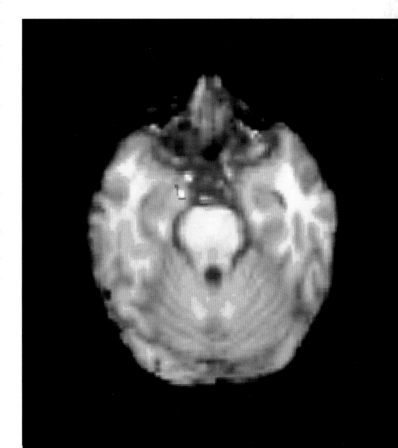

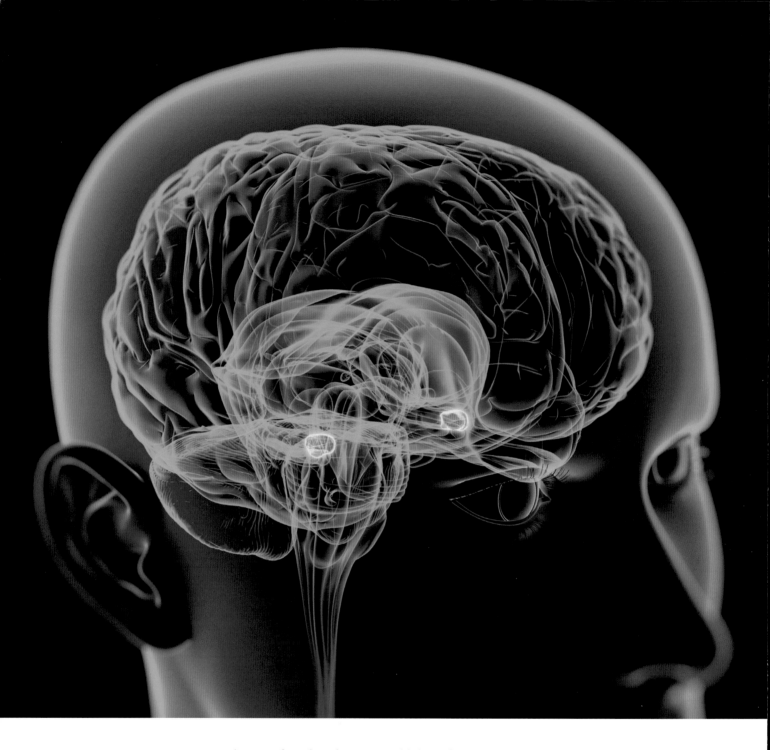

This computer artwork of a person's head shows the brain inside. The highlighted area shows the two amygdala. They are part of the brain's limbic system, a primitive part of the brain involved in emotions, learning and the formation of memories — particularly the memory of feelings.

brain is found in the temporal lobes of almost every complex vertebrate on earth and it seems that when it is removed, the impact on our ability to feel fear is almost absolute.

In over 15 years of extensive studies, it's become clear that SM cannot experience the feeling of fear regardless of the stimulus that is put in front of her. Early studies of her revealed that she was unable to recognise the emotion of fear in other people's faces, and testimony from her life has shown that despite being exposed to multiple traumatic and life-threatening situations, such as being held at knifepoint and gunpoint, she did not react with the panic that such situations would normally be expected to induce. In the laboratory, studies have tested her responses to a range of different stimuli including horror films, facial expressions and even a visit to America's most haunted house, and once again have revealed that her response to threat remains muted and often non-existent. Researchers report that she is of an almost persistently positive state of mind regardless of what life throws at her and is often socially oblivious to the awkwardness of a situation, perhaps a side effect of the lack of social fear that lurks in many of us.

Of course SM is only one case and so extrapolating widely from a single study must remain in context, but after 17 years of investigation there is no doubt that she does provide a deep and intriguing insight into the role of the amygdala in the generation of fear.

# FIGHT OR FLIGHT

Fear is instinctive, an innate reaction that is triggered before you even have time to think. When we are exposed to a threatening or scary situation, the response is automatic and almost always the same and it all begins in the amygdala, the very part of the brain that is so damaged in the case of SM.

All of our senses, but particularly our eyes and ears, are continually monitoring the world for any sense of threat, sending information to many different regions of the brain including the amygdala. As part of the limbic system the amygdala is at the heart of our emotional experience, monitoring the sensory input before we have even had time to consciously perceive it. If the input into the amygdala suggests some kind of threat, then it immediately responds, so fast that it triggers the body to react even before it is conscious of it.

The resulting neurological cascade makes its way through the control centre of the brain, the hypothalamus, which in turn sends signals to the adrenal glands on the kidneys, activating the release of adrenalin that floods the body and causes a wide range of physical effects that we know as the flight or fight response.

The result is almost instantaneous, as the pacemaker cells in the heart wall are activated, making the heart beat faster and so forcing blood rapidly around the body. Adrenalin also attaches to cell receptors in the lungs, expanding the airways, increasing breathing rate and so allowing more oxygen to enter the body and increase the alertness of the brain and senses. In the liver adrenalin dilates

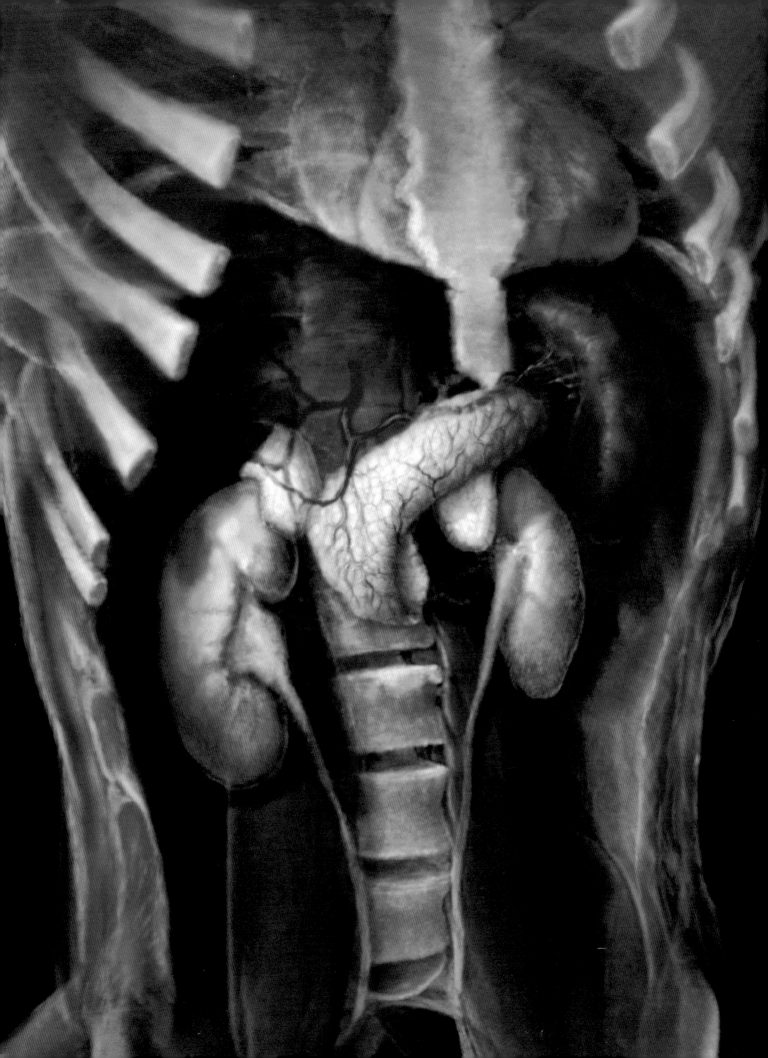

the blood vessels, releasing stores of glucose into the blood essential for powering muscles. With the heart beating faster and the blood saturated in oxygen and glucose, our body is primed for escape.

All of this happens so quickly we aren't aware of it. Bypassing the conscious parts of the brain means that the speed of response is so fast the cascade has started before our visual cortex has even had a chance to process what's happening.

Shortly after the adrenalin has surged through our bodies, the second part of the fight flight response kicks into action with the release of cortisol, the stress hormone. Cortisol triggers an increase in blood pressure, blood glucose and the suppression of the immune system, effectively keeping the gas pedal pressed down on the stress response in the face of continued danger. Only when the threat is perceived to diminish is the brake applied, with a parallel response known as the rest and digest response being activated by the release of a neurotransmitter called acetylcholine. In this way the homeostatic balance of the body is kept in order and prevents the stress response from flying out of control.

**Opposite page:** Three-dimensional visualisation of endocrine glands reconstructed from scanned human data. The adrenal glands, perched on top of the kidneys, serve essential endocrine functions.

Acetylcholine crystals, polarised light micrograph. Acetylcholine is a neurotransmitter that plays an important role in relaying impulses at muscle-nerve junctions.

The system is incredibly efficient and fast but to be that quick means it is prone to false alarms which in turn means we are vulnerable to a fight or flight response even when the threat is minimal or not even there. To control these false alarms requires a different part of our brain, the pre-frontal cortex, where our conscious thought resides, to try and rein in the flight instinct and keep our panic under control.

Our ability to comprehend the brain at any level is so remarkable that we have, perhaps, all been a little bit seduced into thinking we are on the cusp of a total understanding. But measurement of brain activity is, in some ways, still quite crude. Using giant magnets scientists can measure blood flow fairly precisely in space but not in time. They can tell you which part of the brain is getting more blood but not down to the millisecond level that the brain's circuits work at. You can measure electrical activity precisely in time but not in space using scalp electrodes (an electroencephalogram or EEG). These will tell you about electrical activity but don't allow you to locate it very precisely. You can get very, very specific about which neuron is firing if you use fine electrodes during neurosurgery or on animals, but these techniques are unsurprisingly not popular with large groups of healthy volunteers.

So, our understanding of the brain has been pieced together from animal studies, imaging, and mainly by studying patients with deficits – if you know the part of the brain affected by a stroke and the deficits of the patient, then you can get a very clear idea of what functions that part of the brain normally performs. All of this has allowed some remarkable advances, but much of recent neuroscience has been about generating laboratory proof of things that have been experientially known to just about everyone who has ever lived. Most of us know, for example, that the experience of being in love is different to the experience of worrying about our tax returns and we would expect that the activity in various brain regions might look rather different in those two states. That neuroscientists are now able to prove this seems at best unilluminating. Identifying the physical correlates of consciousness can feel like mere re-description but with fancier vocabulary. Saying that the ventral striatum of the basal ganglia lights up when you see something you desire does not in itself tell us very much new about the experience of wanting a cool beer on a hot day.

So, perhaps I wasn't expecting too much from my visit to the Karolinska Institute in Stockholm to meet neuropsychologist Dr Frederick Ahs. He studies fear and how the brain responds to fearful stimuli. Neuroanatomy fades from memory quicker than much of the knowledge acquired at medical school but I expected him to mention the amygdala and probably some other deep-brain struc-

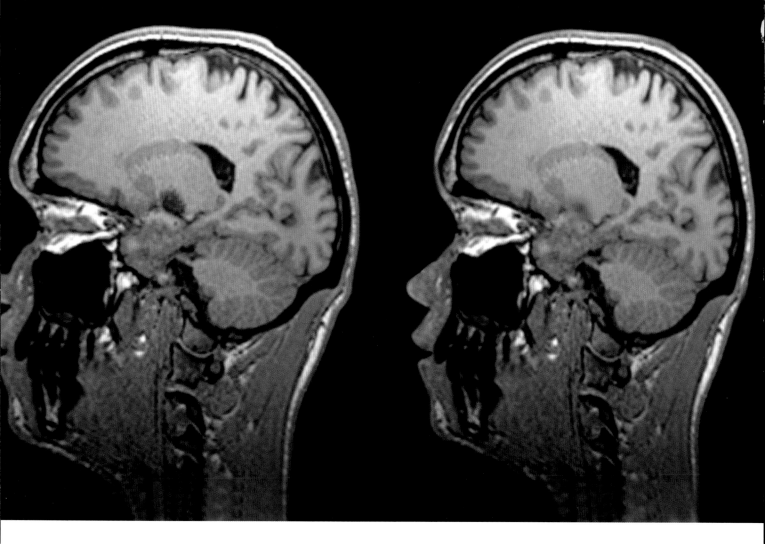

tures I hadn't heard of. I wasn't expecting him to transform the way I thought about the brain and its origins, but this is exactly what he did.

We started the day in a magnetic resonance imaging scanner or MRI. This is a huge magnetic tube. Exactly how it scans the brain defies simple explanation but here goes. With a powerful enough magnet you can cause water molecules in the brain to emit radio waves. When detected they tell you how much water is in each part of the brain. This allows you to identify different structures and also to quantify blood flow. I told you it wasn't simple.

I was slid into the magnet with my head clamped still and undertook the same test that he had given a series of volunteers for a study. I was shown a series of pictures of snakes and mushrooms while my brain was repeatedly scanned for changes in blood flow to different areas. I'm not snake phobic but my brain response could then be compared to people who were snake phobics to see how their brains responded differently. (As a side note: why mushrooms? Because when you show someone a picture of anything their brain lights up, so Frederick needed to control the experiment with some other pictures to reduce the noise in the data. Mushrooms are in some ways like snakes – found in woods, occasionally poisonous – and crucially they've been used in psychological studies before as a control

Conceptual composite MRI brain scans demonstrating differences in jealousy between the sexes. At left is a man's response and at right is a woman's response (superimposed onto copies of the same brain scan). The hypothalamus is shown in red, with the woman's hypothalamus less active. Research has shown that the amygdala (controlling fear and aggression) and the hypothalamus (sexuality) are activated differently in men and women.

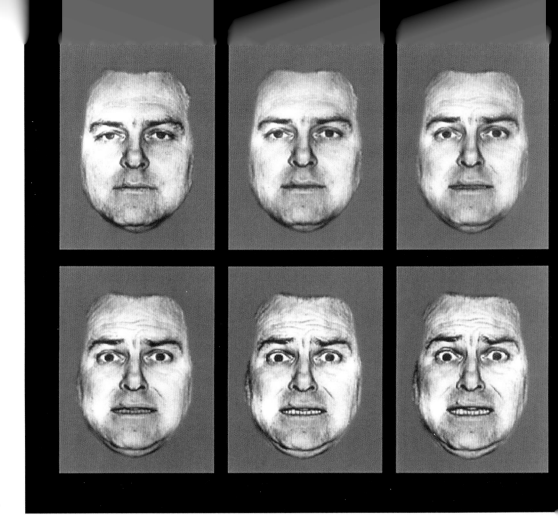

Six images of a man's face showing progressive stages of fear in his expression. At top left is a neutral face; at bottom centre is a prototype fear expression. From the neutral face, fear is progressively added by 25 per cent (top centre), 50 per cent (top right) and 75 per cent (bottom left). Enhanced fear 25 per cent beyond the prototype is seen at bottom right. These faces were used in a study to test the response of the left amygdala in the brain to facial expressions of fear. With increased fearful expression the amygdala shows increased activity, suggesting this part of the brain recognises and responds to facial fear.

(they enabled Frederick to make sure that both groups of volunteers responded normally to non-phobic visual stimuli).

I dutifully looked at the pictures of the snakes, trying hard not to doze off in the warm cocoon of the scanner. When I emerged Frederick showed me composite images from the dozens of brains put through the same test. Yes, the amygdala 'lights up'. The amygdala is popularly understood as an ancient part of the brain believed to have evolved over 200 million years ago. It is intimately involved with the processing of emotion, particularly fear. It is ancient and rapid. In a classic series of experiments two scientists at the University of Uppsala in Sweden showed that fear does not require conscious recognition of the fearful stimulus. By showing phobic subjects pictures of snakes or spiders so rapidly they were not consciously processed, nonetheless they were able to measure fear reactions in the volunteers by increased skin conductance – one of the first things that happens when you are afraid is you start to sweat; this is the principle of the lie detector test.

So, unsurprisingly, the amygdala shows increased blood flow to phobic stimuli but, when you show non-phobic volunteers pictures of snakes, you see another part of the brain light up after the amygdala – part of the pre-frontal cortex. The pre-frontal cortex has a vast range of different functions and occupies almost a third of the surface of the brain. But it is best known for its function in cognitive and emotional control. Frederick describes its role here as the rider of a horse. 'You could think of the pre-frontal cortex as being the rider and the amygdala as being the horse. Sometimes the horse gets spooked but sometimes the rider is still able to control the horse.'

So, this is deeper than mere anatomical description. Here we see communication between two different brain areas. What we're seeing is bravery. The primitive, ancient part of the brain being overridden by the more sophisticated higher centres, those parts of the brain that are uniquely human: no other animal has such a developed pre-frontal cortex. And with that thought Frederick catalyses a deeper discussion about brain evolution and function with parallels throughout the body's systems. 'Much of the higher brain is inhibitory, it has evolved to damp down the simpler responses of the lower brain. These have immense use for speed but higher control has become more and more important as we evolved.'

Modulation of fear is a good example. As with disgust, clotting, immunity, almost all the body's processes and emotions, there is a 'Goldilocks' amount. Too little fear is as harmful as too much. And like all these processes, there are ancient systems being finely regulated by more recent human adaptations.

Understood through the lens of evolution, the system imaged by Frederick's team makes sense. An ancient rapid response with a separate module added later on for fine control. Like many other aspects of our physiology, our fear responses seem out of place in the modern world. For the most part they are underused. Months may pass between moments when I am truly afraid and my amygdala gets a good work out – perhaps there is an immune analogy here. Just as an under-challenged immune system may lead to autoimmune disease, so an unstimulated fear system may allow the development of phobias. Or it may be that these days the stakes are simply not high enough. Throughout history it's likely that most vertebrate males have left no offspring – bravery, at least to a degree, is a good way to increase the chances. After all, dying in a fight for a mate is the evolutionary equivalent of not fighting at all. Either way you leave no genes.

There is a definitive neuroscience textbook commonly known as 'Kandel and Schwartz' (actual title, *Principles of Neural Science*). Eric Kandel, one of the authors, won the 2000 Nobel Prize in medicine for research on how memory is stored. In my edition of the book there is just one single line on brain evolution, indicative that this is not the way the brain is considered often enough. In 12 hours spent with Frederick he illustrated an organising principle and related it to ancient and modern behaviours, providing me with a new way of understanding my own brain and how it relates to the rest of my body.

# THRILL-SEEKING:
# A LOVE OF FEAR ...

There's one final, rather unexpected side to our relationship with fear – we all quite often like it! The modern world offers a myriad of ways of generating a sense of fear in us for nothing more than pleasure. Whether it's horror films, roller coasters or bungee jumps, the consumer market peddles us fear on tap and we are very willing to pay for it, but why?

Our modern-day addiction to thrill-seeking actually has deep roots in our evolutionary past. It's a survival instinct that's been hijacked for kicks from a time when pretty much the only thrill we got was in the hunt for large and potentially dangerous animals. Back in those hunter-gatherer days, it made evolutionary sense to reward the risk of the big catch, because if we bagged it, it would have been beneficial for our survival.

Today we might not be bringing down a bison to feed the village but when we step onto that roller coaster or turn on the film, exactly the same biology plays out. When we are in a fearful situation but realise that it's not going to tip over into immediate mortal danger, we release feel-good chemicals, like serotonin and oxtyocin, which go flooding through the brain. It's a reward for taking a risk and surviving, a neurological pat on the back that over time would have helped us develop better-adapted behaviours as a hunting species.

By reinforcing successful behaviours, this counterintuitive response to fear made us more likely to survive and in doing so seems to have resulted in genes connected with increased thrill-seeking being selected across the generations. Evolution selects blindly but if the gene increases survival, as in the case of thrill-seeking, then its occurrence will only increase. Recent studies have revealed that even today a tendency to thrill-seeking seems to be carried in your genes. For some people this goes too far and the addiction to risk really does push them to the edge and beyond, but for most of us that healthy urge to push ourselves into controlled danger is one of life's true, well ... thrills.

**Opposite page:** A terrifying drop on a roller coaster at Tokyo Dome City Attractions. Our modern-day addiction to thrill-seeking actually has deep roots in our evolutionary past. It's a survival instinct that's been hijacked for kicks from a time when the only thrill we got was in the hunt for large and potentially dangerous animals.

# THE YUCK FACTOR

As we've seen, fear is there to keep threats we can sense and see at a distance and keep us out of danger. We also now think that our ability to control fear and the adrenalin rush we experience when it descends have driven our brains, and ultimately our behaviour, through the process of evolution by natural selection.

There is, however, another emotional response that's central to our survival, one that we develop in the first few months of life and that then develops with us across our lifetime. It's an emotion that is designed to keep a very different sort of threat at arm's length, the type of threat that comes from rancid food or putrid water, a threat that we cannot see but one that we are made incredibly aware of through the reaction we know as disgust.

Professor Valerie Curtis is the author of a book entitled *Don't Look, Don't Touch, Don't Eat: The Science Behind Revulsion*. This book is a fascinating tour through an extraordinary variety of human experience. Well written by someone who is an acknowledged world expert in the subject, it covers sex, murder, food and generally what it means to be human. I imagined that it would have sold extremely well to the vast numbers of people who are curious about science and their bodies. 'It didn't sell at all well,' said Curtis. 'I of all people should have known that the topic would put people off. People don't want to even think about being disgusted.' She said this at the beginning of an afternoon, the events of which would make that point extremely effectively.

Chris and I had invited Val Curtis over for Sunday lunch to eat a meal of disgusting food. Our reasoning was this: disgusting stuff can be rendered completely safe and surely it would be more interesting and fun to explore the value of revulsion while we were actually experiencing it. It turns out to be quite easy to find someone to cater for a lunch where the food does not have to taste good at all: a former colleague who had done the cooking on a diet TV programme I once made agreed to source the ingredients for and safely prepare a disgusting lunch. By the time Valerie arrived the table was set and it contained the following: lamb brains, sheep eyeballs, bull testicles, a roasted pig head, several different kinds of dried insects and maggots, thousand-year-old eggs, and a huge, stinking Burgundian cheese. Most of it had been poached to render it safe and was then unceremoniously dumped into bowls without seasoning or sauce. The kitchen smelled damp and sulphurous with a hint of barnyard. The contents of most of the bowls were grey, with congealing fat and connective tissue coating everything.

None of these foods are universally regarded as disgusting. I quite enjoy a roasted pig head and there is no cheese on earth I won't enjoy (including the Sardinian one with maggots, before you ask – I know it well and I like it). Both Chris and I could sincerely claim to eat a wide variety of different foods. Chris served tripe at his wedding. But while what we were serving that Sunday afternoon was edible and could be regarded as food, it had been prepared with no love and care beyond making sure it was safe to eat, and it all looked and smelled appalling as a collection. There are few cultures on earth that regularly eat unseasoned boiled offal and the ones that do are rarely rewarded by a visit from the Michelin inspectors.

We had invited a couple of other friends to the lunch to make the atmosphere as convivial as possible, and so we all sat down and tucked in. Well, we tried to tuck in. The first challenge was the eyeballs. It is disconcerting to see your food staring back at you. Our friend Imahn cut one up and tried a piece and so, not to be outdone, Chris popped one in his mouth. A sheep's eyeball is slightly smaller than a squash ball but approximately the same texture, except that it has a large bit of gristle at the back (the optic nerve) and it is full of slime (the vitreous humour). As soon as he bit into it, he looked like he would vomit, he actually seemed to be choking back vomit. It was striking that I, and everyone around the table, also began to writhe and gag in the same way. The experience was not only Chris's, it was shared almost equally among all of us. We all had a go at different dishes. The insects weren't too bad but the testicles were appalling. For a start they conjured up extremely vivid images of, well, testicles and when I cut into them they oozed fluid, which I suppose was just residual cooking water but was an unwelcome reminder of their original job in the bull. The really unpleasant part occurred once I had cut into one: the tough outer sac which covers the testicle (within the scrotum) is called the *tunica vaginalis*. This sac is lined on the inside with undulating blood vessels which look exactly like worms. I managed a tiny mouthful and felt instantly sick. So did everyone watching me. And yet despite the

Chris eating honeycomb in the Republic of Congo. Complete with bee larvae.

The owner of the Casa de Oro restaurant in La Paz, Bolivia, serves *Cardan* – a soup made with phallus and testicles of bulls. The soup supposedly has aphrodisiac qualities among other powers.

extreme unpleasantness of the experience, there was something thrilling about it as well. Perhaps the same as going on a roller coaster together, we had bonded over this meal and had spent much of the dinner roaring with nervous laughter as we each tried different things that threatened to make us sick without actually making us ill. It was the culinary equivalent of watching a horror movie, including shouts of 'don't go in there!'

Most of the dishes were all parts of the animal that were unlikely to be eaten in other circumstances, so we didn't feel too guilty scraping much of what had been prepared into the bin. But it was striking that we had all turned down a table groaning with nutritious food because of an irrational emotional response to its appearance or to the idea of it. Why would we have done this? Why do we have such a finely tuned sense of what is disgusting and what isn't?

Curtis describes disgust as

a voice in our heads ... the voice of our ancestors telling us to avoid infectious disease and social parasites. The voice of emotion is there for a reason – it guides us to behave in ways that are good for our genes or, more precisely, to behave in ways that were good for the genes of our ancestors.

 The ancestors to whom she refers are very old indeed, preceding the evolution of our executive brain which, as it does with many primitive emotions including fear, can absorb reasoned argument, consider a variety of outcomes and learn from experience and science.

Her central theory about avoiding infectious disease was certainly supported by my extreme dislike of eating the wormy-looking interiors of the testicles. She has made a list of things that cross-culturally and almost universally elicit disgust. All of them are potential sources of infection. Faeces, dirty water and contaminated foods which present a danger of fecal-oral disease; skin lesions which can spread pathogens through skin contact; ulcerated genitals and a revulsion at individuals such as sex workers, who may present risks of sexually transmitted diseases; sputum, snot and used tissues, which all risk respiratory infections; and finally, in general, sick people and bodily secretions and excretions (the general ooze and pus and sweat of illness), which are the source of multiple possible infections.

There is evidence that our disgust system is not merely an evolutionary hangover. Even in today's more hygienic environment, it seems to be effective: those with lower disgust sensitivity suffer from more infectious disease; there is evidence that, when shown images that are disgusting (as opposed to merely upsetting), people upregulate their immune systems.

Neither Professor Curtis nor I are suggesting that you should be revolted by sex workers or ill people. But these reactions from an evolutionary point of view are interesting precisely because they have a judgemental, moral component. What began as an instinctive device to steer us away from dangerous food or sex has become a moral device to build a society without rampant disease. This is extremely important for humans, as we are 'ultra social' and therefore face a conundrum: we have to balance the benefits of sociality with the increased risk of infectious diseases that being social brings.

Parasites, viruses and other infections adapt to exploit our close proximity and personal interactions, and so feelings of disgust have characterised the emotional responses of what can – in some ways – be considered disease avoidance strategies: (to quote Curtis)

preferring to mix with insiders (ethnocentrism), avoiding outsiders (xenophobia), excluding any individuals that show signs of infection (shunning) or punishing those that behave in ways that may threaten others with disease, by displaying poor hygiene, for example. So as not to be punished or excluded, individuals self-police their own

hygiene and social contact behaviour, sometimes turning disgust on themselves (shame). Group norms of hygiene behaviour (manners) may emerge and groups may agree to cooperate on activities that protect the group as a whole (public health). Because disgust is 'strong magic' that recognises an ability to contaminate by association, it is used to marginalize outsiders to groups (stigmatisation) and is employed in ritual and religion to demarcate what is pure and what is polluted.

This is an extraordinarily wide-ranging and influential collection of behaviours, many of which my progressive politics find, in themselves, repugnant. Of course Curtis is not endorsing any of these behaviours but rather offering a new and important explanation for their existence and persistence. Without disgust we would struggle to live in cities, we would struggle to agree on manners, we would struggle to disincentivise cheating, stealing and other kinds of social parasitism. Moral disgust is essential to the ability to cooperate on a mass scale.

At the end of our disgusting lunch, we had all become disgusting to each other and the foods that some of us might have enjoyed under other circumstances had become disgusting too. Someone did take the cheese home but I didn't want it. Humans are able to track possible infections with our imaginations. This allowed us to stay at home during the plague (which probably didn't help much); it also allows us to 'see' contamination and track the spread of potential pathogens. We scrubbed the kitchen with bleach: an utterly irrational end to a meal that had been prepared as safely as any I had eaten.

There is one time in life when most of us will engage in something that, under different circumstances, would be as repugnant as it is possible to imagine. Sex and the range of stuff we do in bed to the people we fancy. Love is the opposite of disgust. The things you most want to do with the people you fancy are literally the things you least want to do to the people you don't fancy. In evolutionary terms a fear of sleeping with the wrong person is valuable for genetic survival because many sexually transmitted diseases can render us infertile. Much of foreplay may be simply an inspection to figure out how disgusting someone is. Interestingly arousal reduces sensitivity to disgust. It is possibly because of the tremendous opportunity that sex affords for the transmission of disease – because it involves the exchange of body fluids – that it is so related to our feelings about food. The bull's testicles or the thought of eating ovaries are particularly disgusting because they involve sexual organs.

It's not a coincidence: we have evolved to avoid most sexual encounters. Val Curtis quotes Stephen Fry and there can be no more elegant justification for celibacy than this:

**Opposite page:** Blowfly larvae or simple maggots. Likely from the Bluebottle Fly which rejoices in the taxonomic name *Calliphora vomitoria*.

**Opposite page:** Coloured scanning electron micrograph of a T-lymphocyte blood cell (green) infected with HIV (red). The surface of the T-cell has a lumpy appearance with large irregular surface protrusions. Smaller spherical structures on the cell surface are HIV virus particles budding away from the cell membrane. Depletion of the number of T-cells in the blood is the main reason for the destruction of the immune system in AIDS.

I would be greatly in the debt of the man who could tell me what would ever be appealing about those damp, dark, foul-smelling and revoltingly tufted areas of the body that constitute the main dishes in the banquet of love ... Once under the influence of the drugs supplied by one's own body, there is no limit to the indignities, indecencies and bestialities to which the most usually rational and graceful of us will sink.

Disgustability is a trade-off. Too much can render a person functionally infertile. In one study 24 per cent of patients with untreated OCD were virgins and another 9 per cent had not been sexually active for years. And yet selective sexual partner choice is desperately important to avoid sexually transmitted infections. There is a trade-off between good manners and snobbery or bigotry, all disgust related. And a trade-off between healthy behaviour and fastidiousness. You can be too disgusted to share a water fountain with other people and you'll just go thirsty, you won't stay healthy. In situations of starvation you need to be able to expand the range of what you're prepared to eat. So our disgust sensitivity, which is usually dialled up full volume so that we don't cram our mouths with rotting meat every time we get peckish, has to be trainable and to be dialled back under specific circumstances. It is ironic that the study of disgust, which should be a central consideration in such diverse areas of research as public health and jurisprudence, instead is neglected precisely because of the importance of its function. No one likes to talk about it. Yuck.

# UNDER ATTACK

Nothing in biology is perfect and despite all of its efforts to keep threat and infection at arm's length, every human body inevitably succumbs to the pathogenic onslaught every one of us undergoes each day. Every minute of your life you are drenched in the endless drizzle of millions of pathogens – viruses and bacteria that cover every opening in your body that they can find.

Our bodies are the perfect incubator for bacteria. The finely balanced conditions so central to keeping our cells alive also provide the perfect environment for invading organisms. Once inside they can colonise every corner of our bodies and be deadly within the space of a few hours. Just think about how quickly a piece of meat will rot, decompose and decay – that's what bacteria can do to your body if left unstopped.

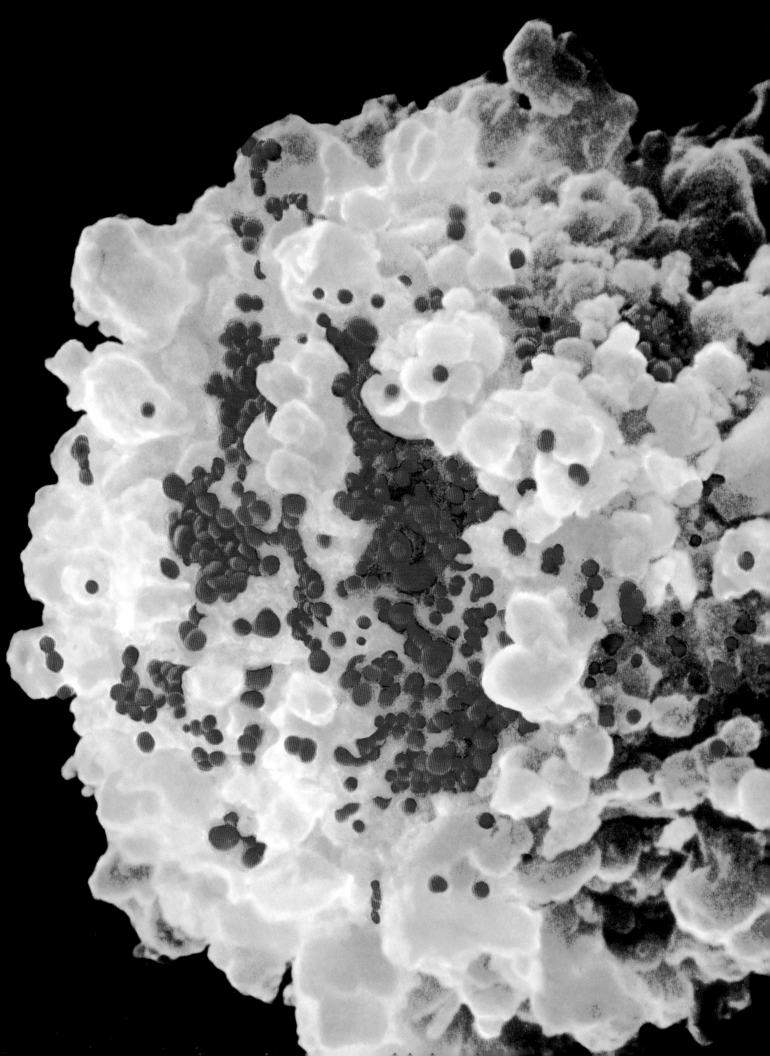

Fortunately for us, we have an extraordinary system of defences. The fluid in our eyes is full of antimicrobial agents, our nostrils are filled with tiny hairs and mucus that act as an infection-fighting trap to grab microbes and stop them from getting in, and our ears are full of wax that contains enzymes which kill bugs on contact. But even these first lines of defence aren't foolproof; there are always pathogens that can get round them and there are millions of them in every breath we take. The most cunning of them are viruses, killers that are up to 100 times smaller than bacteria and that covertly infiltrate the cells of your body, hijacking the machinery within and replicating themselves at frightening speed. Our fight with viral pathogens is a long story of wars won and battles lost, and it's here we turn to the next subject in our story of survival.

# NATURAL BORN KILLERS: NK CELLS

Since the first cases of Ebola over 40 years ago, the number of outbreaks has been relatively small but the mortality rate remains incredibly high, up to 90 per cent in some occurrences. The virus itself, a single strand of RNA (approximately 19,000 nucleotides long), codes for just seven proteins but despite its simplicity its impact remains deadly, rated as the highest Level 4 pathogen by the WHO. Medical care of those infected by the virus continues to be at best hit and miss with no guarantee that even the very highest level of treatment available will be capable of preventing death. At this point the main strategy for controlling an outbreak remains the containment of communities infected and the confinement of individuals – only marginally more advanced than how we dealt with epidemics in the nineteenth century.

The reason Ebola remains such a feared threat is simple – despite 40 years of infection, our primary line of defence, our immune system, is still unable to detect and destroy the virus before it takes a foothold in our bodies. Whilst we slowly endeavour to create an effective vaccine, our bodies remain powerless to protect us from the destructive force of the virus. We know very little about how the Ebola virus evades our immune system so effectively, but what we do know is that this kind of uncontrollable threat is rare. Most of the time our body's defences are more

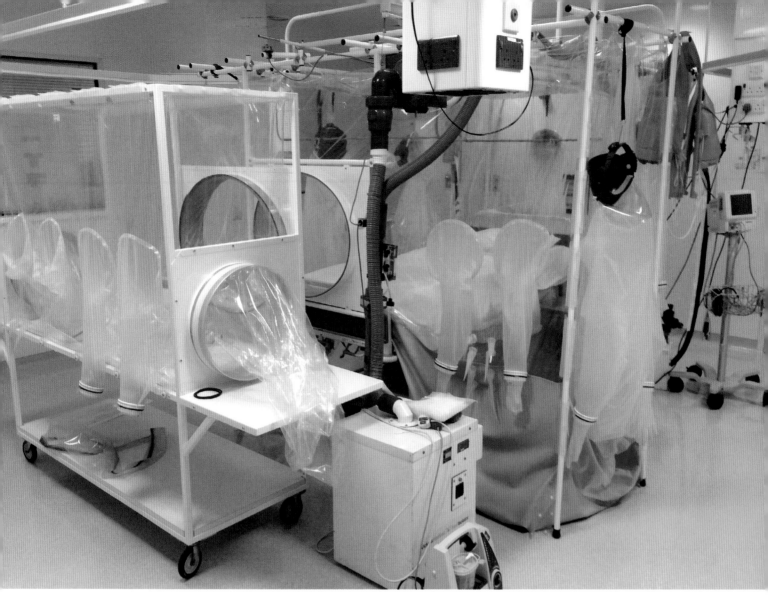

than enough to deter any viral invaders, despite the fact that we are under attack from billions of viral cells each day. To combat diseases like Ebola in the future we need to understand the virus in minute detail and also understand how we can unleash the immune cell that is at the heart of our defences, an extraordinary cell that we are only just beginning to understand.

The swamp forests of central West Africa have dense, thorny foliage. Hunting in these forests lacerates any exposed skin. I know this because as a young doctor I made two trips to these forests and spent months living with the indigenous Bayaka pygmies, and we frequently went hunting. We only caught a monkey once, immediately butchering it on the forest floor. I ended up covered in monkey blood and more certain than ever before about how the global HIV pandemic began.

The origins of HIV have been obscured by rumour, conspiracy and myth. The virus has variously been hypothesised to have been created by the CIA, caused by humans having sex with chimpanzees or spread deliberately by vaccine scientists. In fact we now know almost exactly what happened. Comparison of

The high-level isolation unit (HLIU) at London's Royal Free Hospital is the only unit of its kind in the UK. It is equipped to deal with the most deadly of infectious diseases.

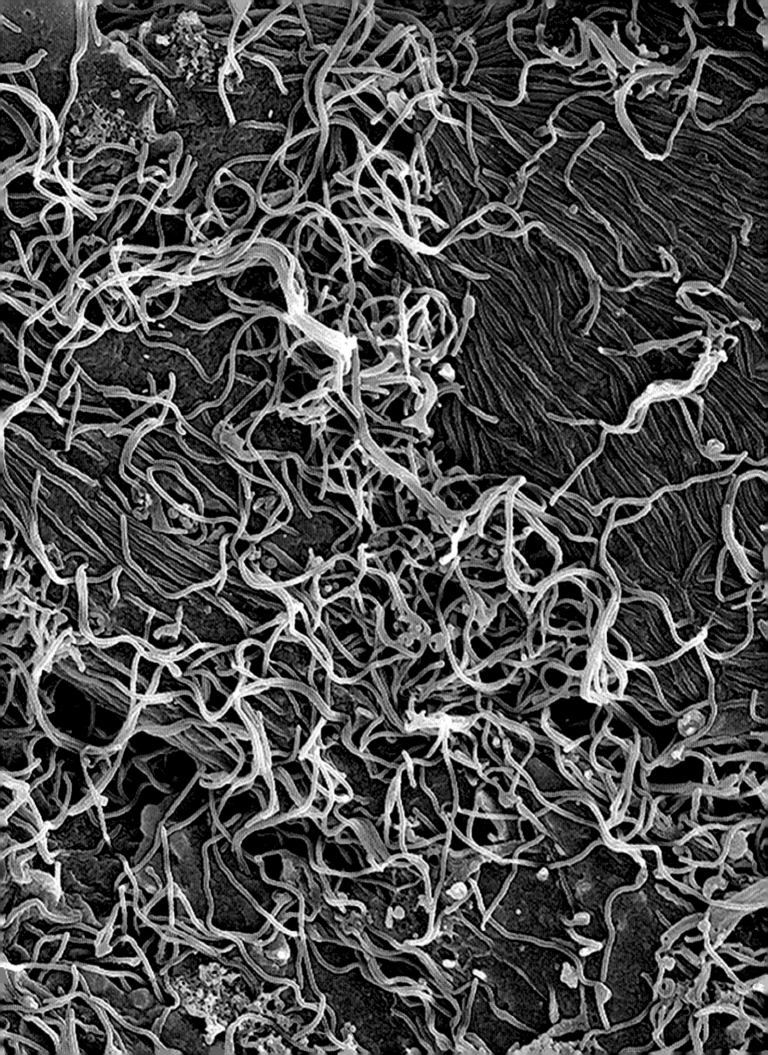

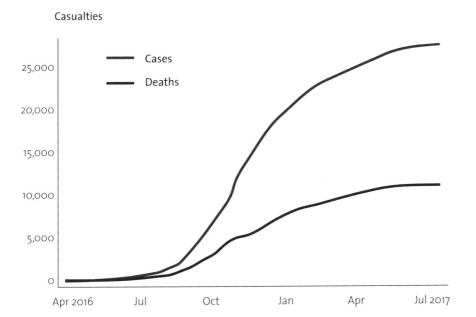

Casualties

— Cases
— Deaths

The development of the 2016–17 Ebola epidemic.

genetic sequences of modern and ape viruses has narrowed down the location when the virus made the jump to a single group of chimpanzees on the border between Cameroon, Central African Republic and the Republic of Congo. This same genetic comparison enables calculation of an approximate date when the virus crossed the species barrier. In the early part of the twentieth century, a 'cut hunter' killed a chimpanzee infected with Simian Immunodeficiency Virus – the ape equivalent of HIV. The virus entered his body through the cut and this single event led to the pandemic that has infected over 60 million people and killed more than half of those.

I was drawn to those dark, steaming forests because of their reputation as the cradle of some of the world's most dangerous diseases. My trips to the forest catalysed my career as an infection doctor and HIV scientist. Over a decade later, at home in London, I came face to face with another West African virus. I am on the team of doctors in the UK who look after patients with the most dangerous diseases, the Category 4 pathogens. These are diseases that are highly infectious, extremely dangerous and usually untreatable. The best known of these is Ebola.

In 2014, a few days after returning from my honeymoon, I got an email from one of my bosses asking if I would be free to work some shifts at the High Level Isolation Unit at the Royal Free, looking after a nurse who had contracted Ebola whilst working in the epidemic that was then rampaging across West Africa. The HLIU is the only one of its kind in the UK and enables high-level treatment of extremely sick patients while keeping them totally isolated from the clinical staff. The patient is kept in a special tent called a trexler isolator. The skin of the tent is transparent and

**Opposite page:** Coloured scanning electron micrograph of Ebola virus particles (red) budding from an infected cell (blue). The Ebola virus disease is a severe and often fatal disease.

there are human-shaped invaginations in its wall so that you can push your entire upper body inside the tent without any physical contact with the patient and without the need for protective body suits. The tent is under negative pressure so that if there is a breach, air is sucked in rather than blown out. And the room in which the tent sits is also isolated with an airlock and under negative pressure compared to the hospital outside. Air can only flow in. Not out. You enter the room from one direction and leave through a shower. Everything that comes into the room with you must be destroyed as you leave, so you have to wear paper underwear.

It's a challenging environment to work in and the patients I have cared for in the tent (one with Ebola and one with another haemorrhagic fever called Lassa fever) have both been extremely sick, but the challenges of working in the HLIU are nothing compared to those my colleagues who went to West Africa faced.

Peculiarly, catching infections from patients is seldom a risk in infectious disease medicine. Patients get infections for complex reasons and most pathogens, whether virus, parasite or bacteria, find it hard to transmit in a clinic. This has not been true for Ebola: over 500 health-care workers died in the most recent epidemic.

Bayaka pygmies live in the Lobaye region, on either side of the border separating the Central African Republic and the Congo.

The story of the Ebola virus and its infection of humans can be traced back perhaps more precisely than any other viral infection in history. On 26 August 1976 Mabalo Lokela, the headmaster of a small village school in northern Zaire, as it was then known, began exhibiting malaria-like symptoms after returning from a trip the previous week to a northerly part of what is now the Democratic Republic of the Congo. His illness rapidly took on a significantly different set of symptoms to malaria, including vomiting and diarrhoea followed by internal and external bleeding, and on 8 September, 14 days after displaying the first signs of illness, he died in the Yambuku mission hospital. Although the Belgian nuns treating Mabalo at the hospital didn't know it at the time, they had been treating one of the very first cases of a human infected with the Ebola virus. Almost immediately those who had come into contact with Lokela began to succumb to a similar set of symptoms as the outbreak started to spread. Over the next 26 days, 318 cases were recorded, with 280 of the individuals dying from the disease, a fatality rate of 88 per cent. This outbreak in Zaire occurred almost simultaneously with an outbreak in Sudan, marking the beginning of this particular type of viral haemorrhagic fever as a human pathogen – we don't know why it struck this time, but we do know it emerged from a viral pool in fruit bats. Studies of the Zairian outbreak would lead Belgian medic Professor Peter Piot and his team from the Institute of Tropical Medicine in Antwerp to name the virus they discovered (in the blood of one of the nurses) after a nearby river – the Ebola River, a name that was chosen to distance the newly identified virus from the town where it was first discovered but would ultimately stigmatise a whole region by its association with the much feared contagion.

Perhaps because of the ground prepared by the experience of the HIV epidemic, the emergence of Ebola simultaneously terrified and fascinated the world. An international team of scientists and doctors made trips to the region and collected epidemiological data along with virus samples to piece together the origins of this epidemic. Ebola is too deadly to have ongoing transmission in human populations like measles or HIV. It lies dormant in an unknown animal reservoir and emerges occasionally, killing a few hundred people. Most epidemics are halted with basic measures like the closing of the local hospital, avoiding ongoing transmission. This latest epidemic was different. The virus spread rapidly across West Africa, killing over 25,000 people.

The first few cases were traced back to a single 2-year-old child in Meliandou village in Guinea, who died after four days of illness characterised by fever, black stools and vomiting. His mother, grandmother and sister all died in the next four weeks.

Detailed investigations of the village revealed a hollow tree that children, including the index case, used to play in. This tree caught fire a few months after

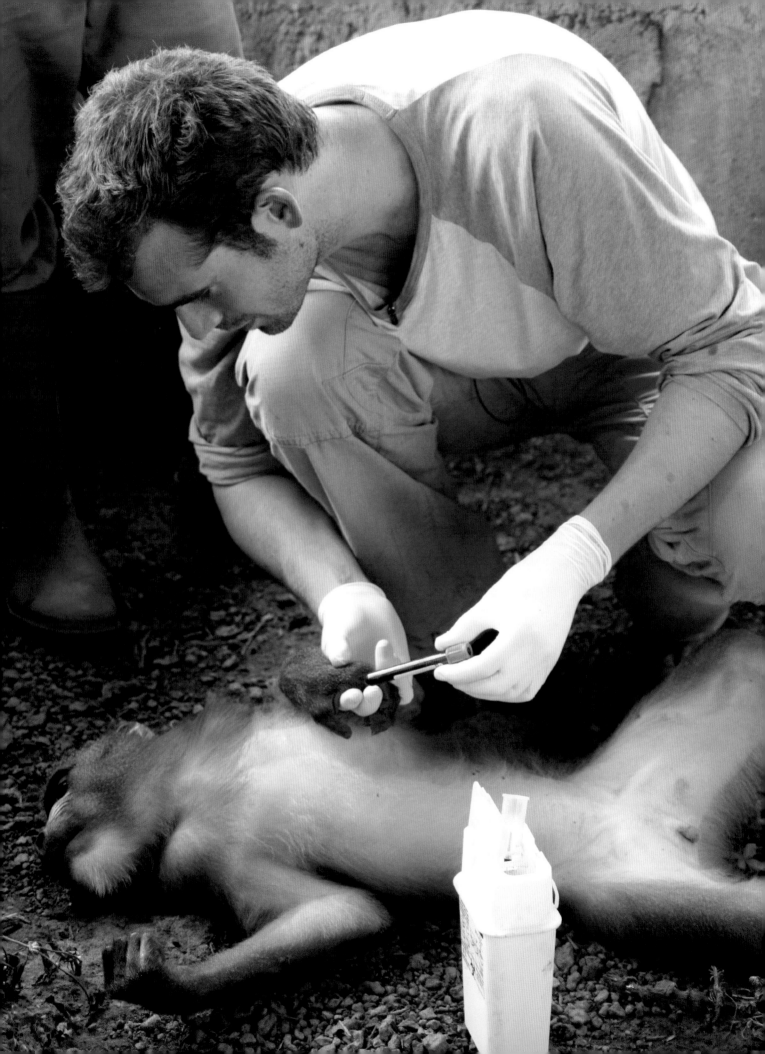

the outbreak of the epidemic and once on fire, a 'rain of bats' fell from the tree. Playing under the tree may have allowed massive exposure to the virus in bat droppings. While bats are believed to be the reservoir, the evidence for this is largely circumstantial. No one has yet been able to obtain samples from bats with a virus that matches the one which spread in the epidemic.

The response from the international community did not match its hysteria. Despite pleading from international NGOs, the UN and governmental response was slow. Western governments responded appropriately only after it became clear that their own populations, not to mention their financial markets, could be at risk. Once the epidemic took hold, it was over a year after the first case that it finally burnt out.

Beyond the gigantic human cost, what is striking about both the Ebola and HIV epidemics is how comparatively unusual they are. Children play under bat-infested trees every day across Africa. Humans have hunted bushmeat in the

**Above:** A village on the banks of the river Ebola.

**Opposite page:** Chris taking a blood sample from a young male Mandrill in Gabon, checking for Simian Immunodeficiency Virus. The virus is so similar to HIV that a human HIV test can be used.

**Opposite page:** Coloured scanning electron micrograph of a natural killer (NK) cell. NK cells are a type of white blood cell and a component of the immune system. They recognise certain proteins, or antigens, on virus-infected or tumour cells and destroy them.

forests of West and Central Africa for millennia. But deadly viruses that jump the species barrier and spread are fantastically rare, despite the fact that every second of every day you are exposed to thousands of viruses largely unknown to science. The amazing thing is that we are able to survive this onslaught at all.

To understand how the human body does this, I went to visit Professor Dan Davis at the University of Manchester. Dan is a physicist turned biologist with an immense enthusiasm and talent for his subject. His speciality is a type of immune cell called a natural killer (NK) cell. These are the brutal secret police of the immune system. They wander the body going from door to door looking for cancers and virally infected cells.

When a cell is infected with a virus it does an ingenious thing. It takes parts of the virus and shows them on its surface attached to molecules called MHC proteins. If immune cells called cytotoxic T-cells recognise fragments of viral proteins attached to MHC proteins on the surface of a cell they know that the cell is infected – and kill it. So, many viruses evolved a strategy where they simply take the MHC molecules off the surface of the cell, preventing the recognition of the virus by the cytotoxic T-cells. Dan explains: 'Viruses act like a gang of criminals pulling the blinds down in their hideaway so no one can see that they're there. But the NK cells have evolved to simply kill cells with the blinds down.'

If the NK cell doesn't get a handshake from an MHC molecule on the surface of one of the body's cells, then it simply assumes criminal (viral) activity within and destroys the cell.

In a further evolutionary twist some viruses like cytomegalovirus have taken to manufacturing fake MHC molecules that sit on the surface of cells and reassure NK cells – rather like putting up a picture in the window of a family eating dinner, reassuring the unobservant secret policeman that there's nothing untoward going on in the house.

Dan's discovery was that NK cells have something more in common with neurons than you might expect from an immune cell. In the 1990s it was noticed that immune cells make prolonged but temporary contacts between each other to exchange information. The cells create rings of proteins in their surface where they contact other cells, through which chemical information can pass. They looked exactly like the synapses between neurons. Dan found that NK cells make similar structures when they contact infected cells and can use them to deliver highly focused blasts of destructive chemicals to infected cells if they don't get the MHC handshake they require.

Dan first shows me a time-lapse movie of a mixture of NK cells together with infected and uninfected cells in a dish. The NK cells move around from cell to

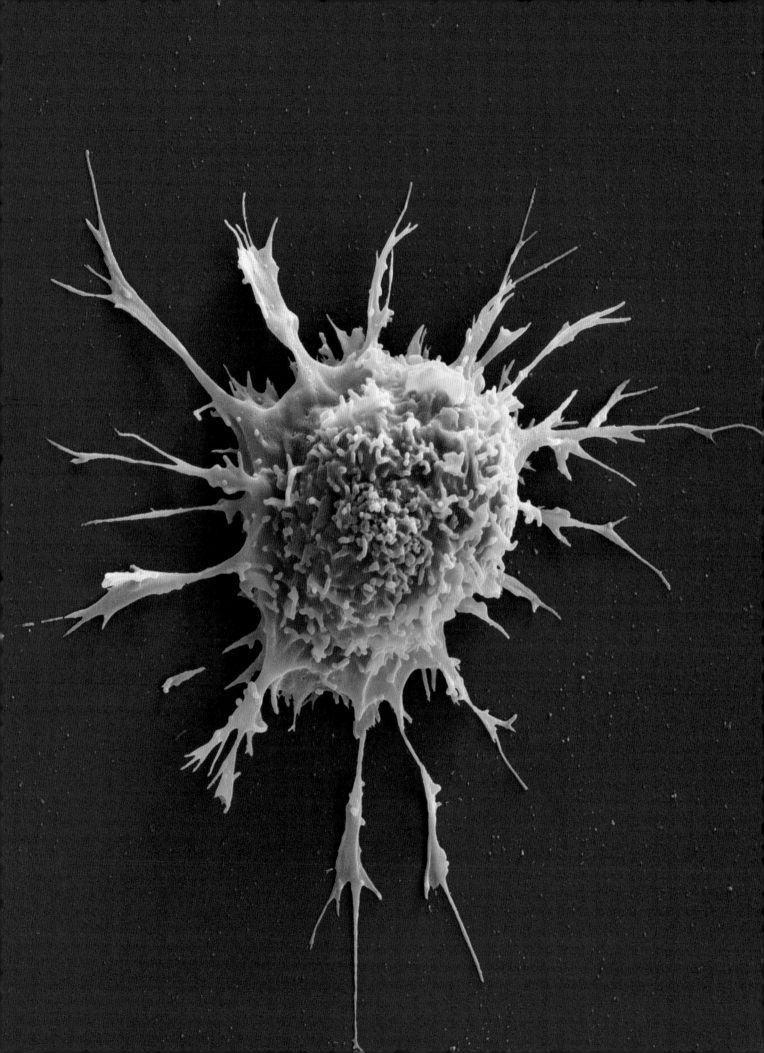

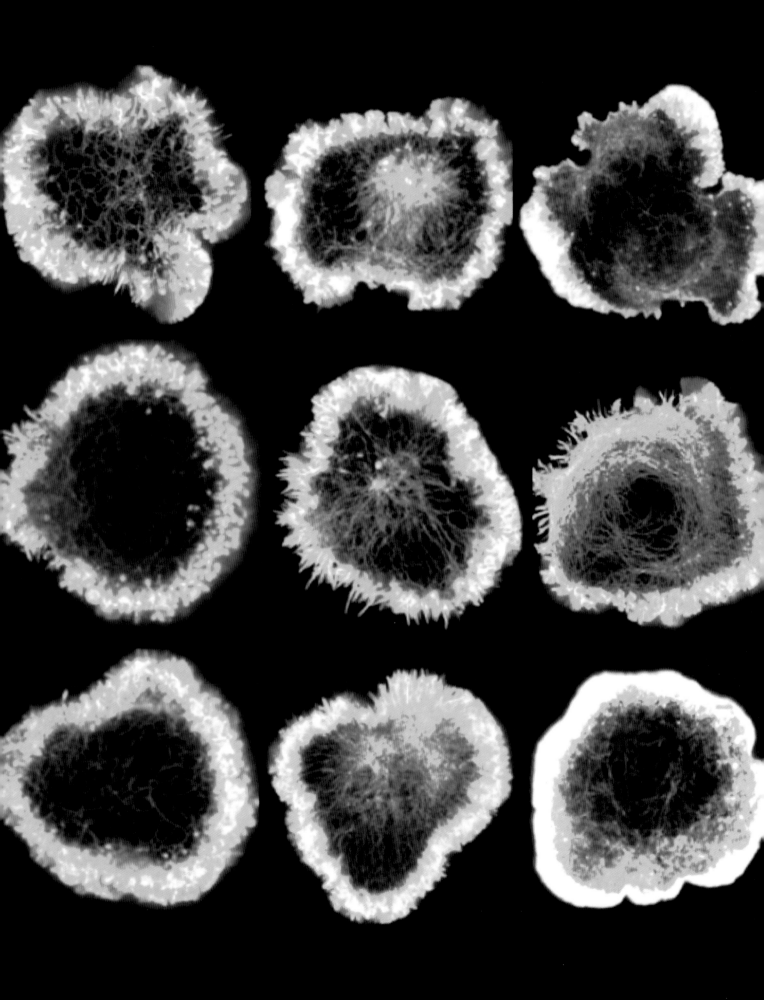

■

'VIRUSES ACT LIKE A GANG OF CRIMINALS
PULLING THE BLINDS DOWN IN THEIR HIDEAWAY
SO NO ONE CAN SEE THAT THEY'RE THERE. BUT
THE NK CELLS HAVE EVOLVED TO SIMPLY KILL
CELLS WITH THE BLINDS DOWN.'
**PROFESSOR DAN DAVIS**

■

cell appearing to kiss each one. When they meet an infected cell there is a more prolonged and ultimately deadly kiss, with the virus-laden cell disintegrating under a barrage of highly focused toxic chemicals. Then he shows me images of these synapses forming. These are taken from the perspective of the infected cell looking at the NK cell using microscopes so sophisticated and powerful they all but defy physics, with a resolution higher than the diffraction limit of light. A terrifying flaming orange ring of proteins assembles on the surface of the NK cell with a central pore, like the gaping mouth of some horrible face-sucking alien. This is a highly organised dynamic structure made possible by precise reorganisation of the protein skeleton of the cell.

It's hard to know where to start with the wonder of it all. Perhaps for me it is with the parallel with the nervous system. Like the nervous system the immune system has memory, it is able to learn from experience. It's dynamic and uses rapid cell-to-cell signals to control a diverse range of responses. And like the nervous system it needs its responses to be finely balanced. Too little and you get infected. Too much and you suffer with autoimmune disease. Finally, like the nervous system it is modular, retaining ancient components, with increasing layers of sophistication added later in evolutionary time for more delicate control. Natural killer cells are one of the first lines of defence. They are crude and brutish but they are capable of activating the next layer up, the adaptive immune system, which will be able to recognise and remember the threat for a better response next time it is encountered.

But, of course, for all its miraculous sophistication the human body never quite gets ahead of the viruses. The formation of synapses both immunological and neurological allows for exquisitely sensitive communication between cells. But it is also a process that can be exploited. HIV and probably other viruses don't like

**Opposite page:** Super-resolution light microscope photographs of actin rings at immune synapses. You are looking from one cell into another through the portal of the synapse. Through the entry of the ring toxic chemicals can be delivered in a highly specific manner, allowing focused cell killing.

hanging around in the blood where they're vulnerable to attack by antibodies and white blood cells. They would much prefer to spread from cell to cell as quickly as possible. HIV infects white blood cells and has evolved to hijack the very synapses that should be halting it to further its own spread. By modifying the cytoskeleton in the same way as in an immunological synapse, the virus forms a virological synapse between two cells which acts as a conduit, allowing for the ferociously fast and efficient spread of viruses perfectly protected from the prying eyes of the extracellular immune system.

The battle is constant and the balance occasionally tips catastrophically against us, but your moment-to-moment survival is a testament to the extraordinary power of your immune system to detect and destroy viral infections before they have even got going.

This is a secret of the human body that until recently was invisible to us – the moment our immune system detects and kills a virus that is hidden within cells in our body. When you breathe in, the lining of the lungs becomes a key battleground in the fight against pathogens. It's here where the invading microorganisms face the full force of our immune system.

White blood cells (like the NK cells) are permanently stationed here in greater numbers than elsewhere in the body, doing their job of detecting infected alien cells, but the speed of the response is a critical factor in preventing the invading pathogen gaining ground. And your immune system has a sophisticated secret weapon in this fight – a memory. A complex system of vessels and nodes called the lymphatic system coordinate the immune system's response to assault. Specialised white blood cells take a sample of the invaders and carry it back to a lymph node, which holds a library of pathogens that have been encountered in the past.

If recognised, this triggers the production of custom-made antibodies designed to target the invader. These flood the bloodstream in their billions and kill the pathogen. It's the immune system at its most effecitve. It not only makes our response to infection faster but it means the more threats the body encounters over a lifetime, the better it gets at killing them.

It's what immunity is.

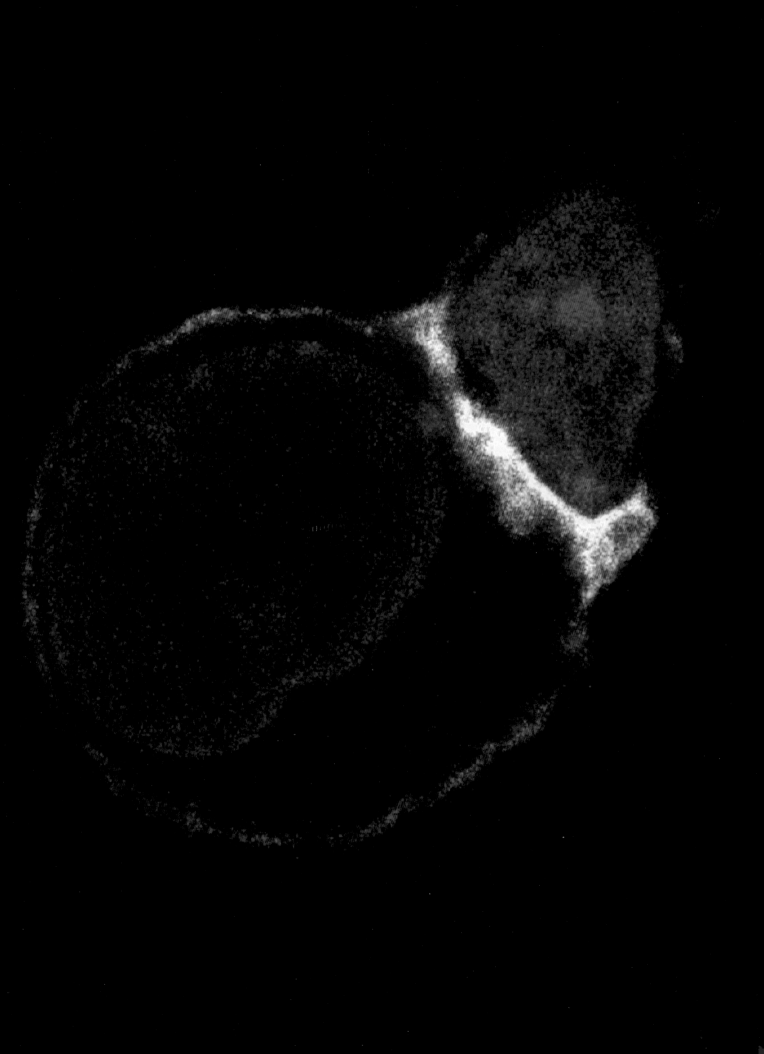

# FIX YOU

As we've seen throughout this chapter, your body goes to great lengths to protect the inner balance that it's maintained every second of every day since the spark of life began at the moment of your conception. To do this we have evolved early warning systems like fear and disgust to try to keep us out of harm's way, and developed an extraordinarily complex immune system to fight off even the most deadly bacterial and viral threats. But we live in a chaotic, dangerous and unpredictable world and that means no matter how many precautions we take, we all ultimately end up getting injured. Thankfully, though, we have developed a remarkable ability to deal with the damage we do to our bodies, whether we decide to walk, run or dance our way through life.

Like every professional dancer Kate Byrne is a master of deception. On the surface she appears the very definition of serenity and grace, moving across the floor with an ease that comes from years of hard work and dedication. It's why she

A dancer prepares her feet before a ballet rehearsal. Every dancer has feet that look more like they have been kickboxing than caressing the boards, as they succumb to a daily bombardment of damage and destruction.

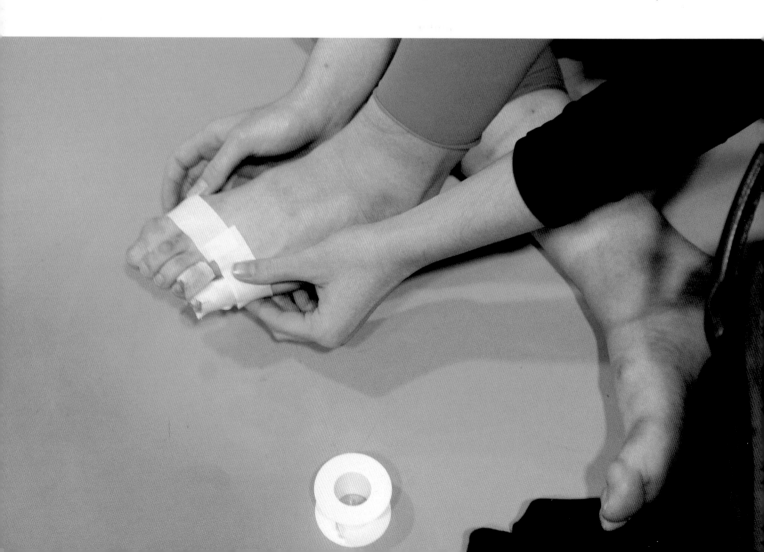

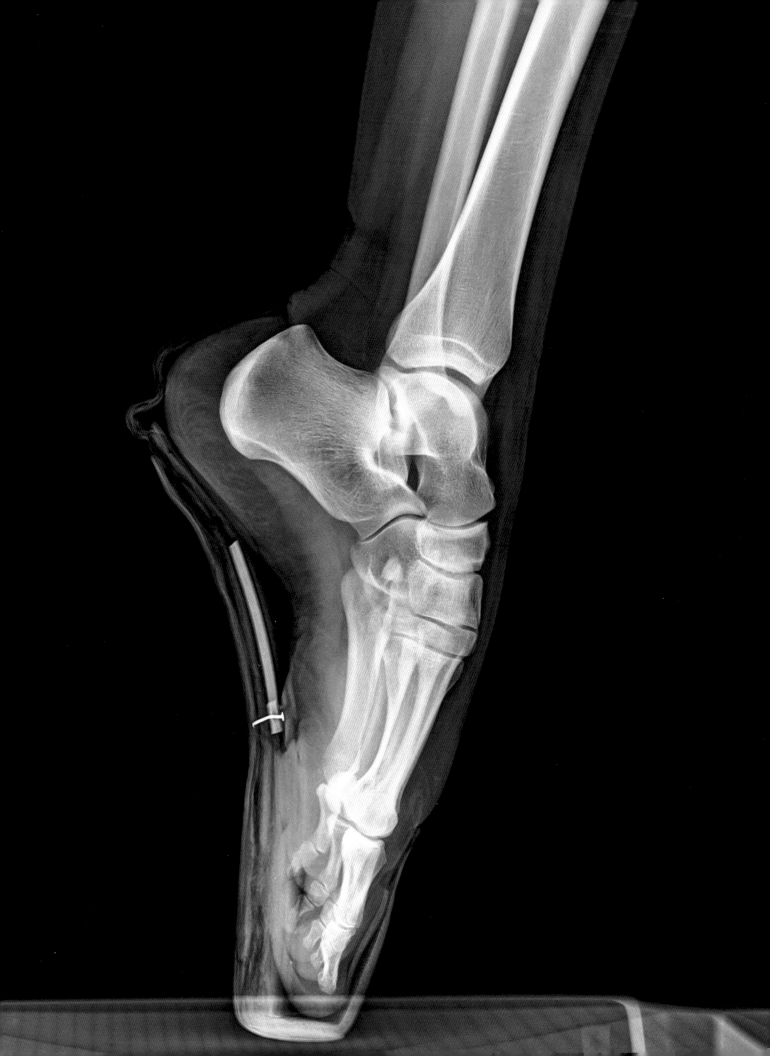

'BALLET IS A MIXTURE OF BEAUTY AND BRUTALITY. I'VE BROKEN TOES, HAD STRESS FRACTURES ... EVERY TIME MY FOOT HIT AGAINST THE FLOOR, THE PAIN WENT RIGHT THROUGH MY LEG AND I COULD FEEL IT IN ALL MY JOINTS, FEEL IT UP MY HIP ALL THE WAY UP MY SHOULDER TO MY NECK ... IT WAS EXCRUCIATING.'

**KATE BYRNE**

has appeared with some of the finest ballet companies in the world and on some of the greatest of stages, from the English National Ballet to the Royal Opera House. Beneath the veneer of calm, however, is a dark and bloody secret, a secret that she shares with every dancer who puts in the gruelling hours of dedication needed to make it to the top.

Peer beneath the satin surface of any ballet slipper and you will see a landscape that is far from beautiful. Every dancer has feet that look more like they have been kickboxing than caressing the boards, as they succumb to a daily bombardment of damage and destruction. It's a barrage that requires constant treatment, from sticking plasters, to ice baths, to the more serious long-term physiotherapy and medical support that professional dancers need to make it through a successful and prolonged career. But no matter what we do to treat our injuries on the surface, it's nothing compared to the power of the astonishing regenerative mechanisms inside our bodies that continually maintain and protect us.

Every time our bodies become injured, from the smallest graze to the deepest cut, a remarkable coordinated response unfolds around the injury that protects and prevents further damage or infection and ultimately restores the area through the process of repair.

Let's take the bleeding toe of our ballet dancer, damaged after a day of punishing practice, and see what happens after the music has stopped. The process of wound healing splits into four key stages – haemostasis (blood clotting), inflammation, proliferation and remodelling.

**Opposite page:** Colour enhanced light micrograph of human blood clotting, with strands of fibrin (red-orange) visible in the plasma.

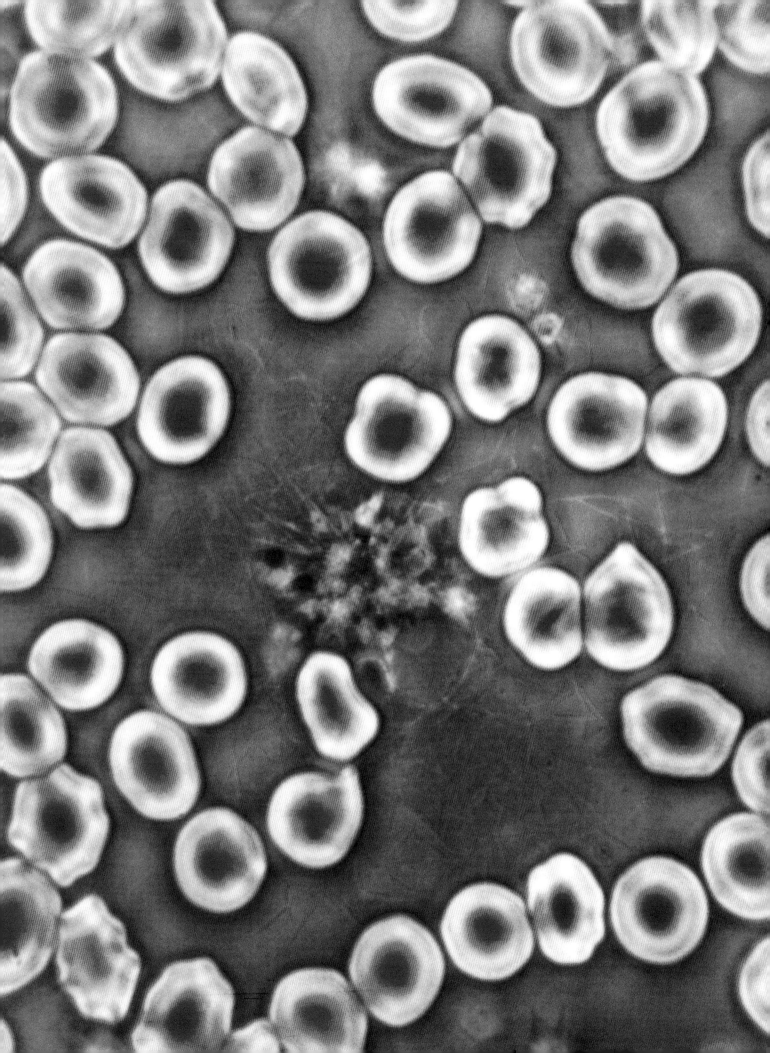

First is the rapid response – the immediate need to protect the body's inner balance requires a reaction within seconds of injury. If even the smallest blood loss is not stopped quickly, then the body's ability to maintain its homeostatic equilibrium would rapidly spiral out of control, a situation we see in the blood-clotting disorder haemophilia. To stop this happening the area around our bleeding ballerina's toe would immediately undergo a series of changes. First the blood vessels constrict, limiting blood flow to the area. At the same time tiny components of blood called platelets begin their crucial function of sticking around the injured site. This instigates a chemical cascade that creates a mesh of fibrin and platelets that act as a plug to slow and prevent further blood loss.

Once the immediate danger of blood loss is under control, the next stage can begin – inflammation. During this stage white blood cells, particularly macrophages, flood into the area to destroy any bacteria that are infiltrating the site; they also remove any damaged or dead cells from the area in preparation for repair. To aid this response another type of white blood cell called a basophil releases the chemical histamine into the area. This causes the blood vessels in the area to become more permeable, allowing the macrophages to leave the blood vessels and get directly into the damaged tissues. The effect of the histamine also causes the blood vessels to dilate and allow plasma in the blood to leak into the surrounding tissues, which results in the characteristic redness and swelling of a wound.

Our sensory network is also activated by the trauma. Nerve fibres known as C fibres in the skin are sensitised by the damage, and control and adjust the flow of chemicals depending on the severity of the injury, creating that sensation we call pain. A warning we cannot ignore, this makes sure we protect the area of damage from further harm.

## WHAT'S THE POINT OF PAIN?

Whenever we stub our toes or get a paper cut, our sense of pain always seems like a cruel trick of evolution. Why couldn't we simply be aware of a sensation and know intellectually to avoid it rather than really having to suffer? As so often with neurology, to understand the function of a sense you need to look at people who don't have it.

Leprosy is one of the best known, least well understood diseases. Despite its reputation it is scarcely infectious. If you have travelled in Asia you will have been exposed to the bacteria which causes it – a bacteria related to tuberculosis: *Mycobacterium leprae*. And if you are reading this book I can be sure almost to the

**Opposite page:** Xand with silver spikes through his face, attempting to see if by entering a religious trance he would become immune to the effects of pain. He didn't.

Chris and Xand feeling exhilarated following a two-hour procession with spikes in their face to the Batu Caves in Kuala Lumpur. Aside from a slight sting eating curry that evening, there were no lasting effects.

point of certainty that you won't have caught it. It is a disease of extreme poverty. I have never worn gloves for the patients with leprosy I have seen at the Hospital for Tropical Diseases.

Patients are widely known to have their peripheral tissues eroded. The bacteria grow slowly at low temperatures of 27–30°C and so they infect the skin and peripheral nerves. The subsequent immune response to these slow-growing bacteria is what mainly damages the nerves. Leprosy is the most common treatable cause of neuropathy (nerve dysfunction) in the world. There are still hundreds of thousands of new cases each year, mainly in subcontinental Asia, although there are around 200 new cases each year in the United States (including a handful infected through contact with armadillos).

Leprosy is seldom fatal – one of the main consequences of infection is as a result of the minor trauma sustained due to lack of pain sensation. Without feelings of pain tissue is rapidly eroded. Throughout the day you are constantly taking small steps to ensure that your tissues don't die or get eroded. Removing the tiny rock in your shoe, rapidly putting down the hot cup of tea. These small actions protect you more than you can imagine from debilitating tissue damage. In a village in the Republic of Congo, Moyamba, where I worked for several months, a man with severe leprosy was tasked with removing the family cooking pot from the fire. Just because he was unable to feel pain didn't mean that his tissues weren't horribly burned and damaged by this daily insult. Similarly, in immobilised patients who are, for example, unable to roll over in their hospital beds, we see the development of bed sores very rapidly. If not treated properly these can prove rapidly fatal. Pain is what protects you from this.

Leprosy also illustrates the strange nature of pain. While peripheral tissues can be numb, simultaneously the patient can suffer with chronic, or long-term, neuropathic pain. This pain is unrelated to painful stimuli but instead seems to originate in the central nervous system, the spine or the brain. This pain has no biological function and is poorly understood but the pathway to the brain seems to be far more sensitive to weak, harmless stimuli which are interpreted as pain.

Pain of the kind we all experience in small amounts each day as a result of a noxious stimulus (stubbing a toe) is carried by two types of nerve fibres, A-delta nerve fibres and C-fibres. The A-delta fibres carry fast messages to the spinal cord while the C-fibres are slower. It has been hypothesised that inputs from these fibres can be interrupted by activation of other sensory inputs (rubbing the stubbed toe, for example). This theory called the 'gate control theory' has transformed our understanding of pain. Although it is now believed to be over-simplified it prompted a vast amount of research with wide-ranging clinical implications.

# THE SMALLEST SURVIVORS

Today we understand the processes behind the human body's fight for survival in more detail than ever before. We have unlocked the intricate web of homeostatic controls that maintain our internal environment, decade after decade, with levels of precision that are simply remarkable. Protecting all of this are inbuilt emotions like fear and disgust that guide us away from danger, an immune system that is primed and prowling 24/7, and a repair mechanism that rebuilds us from the inside out. All of this adds up to a deep understanding of the human

WHOEVER SAVES A LIFE, IT IS CONSIDERED AS IF
HE SAVED AN ENTIRE WORLD
**YERUSHALMI TALMUD 4:9**

body at work, but this knowledge has also allowed us to develop something even greater. In moments of need when the body can no longer protect itself from trauma, illness or disease, medicine can use this understanding to treat the human body with immense precision and skill. From the small things like stitching a wound without leaving a scar, to replacing fluids with a simple sachet of salt when a tummy is upset, to the most extreme interventions modern medicine has to offer. That is what an intensive care unit does – it takes control of all those homeostatic mechanisms, our levels of oxygen, $CO_2$, pH, glucose, water and so much more and maintains them in the face of the most grave circumstance. The more we understand the mechanisms we use to survive, the better we get at helping our body when it's at its most fragile.

Perhaps nowhere is this more apparent than at the Neonatal Intensive Care Unit in St Thomas's Hospital in London. Every day, the teams here apply cutting-edge knowledge to the very smallest lives on the very threshold of survival. In this unit babies born as early as four months premature, 23 weeks into pregnancy, are cared for. Survival in this grey zone of viability is never certain and many of the babies born at this point will not make it through. But despite often being born without the ability to breathe independently, without the ability to digest any food, with unstable hearts and leaking blood vessels, anaemia, immature immune systems incapable of fighting off infection and massively underdeveloped senses, some of these babies can be supported with technology that acts as if they are almost still in the womb, nurturing them until they are capable of independent life.

With the level of technology and support we are able to give to these babies today, it's sometimes difficult to know where we can go next, but medical research is always on the edge of generating its next surprise.

In a research laboratory at the Children's Hospital of Philadelphia, scientists have recently created a system that directly mimics the conditions inside the pregnant mother and have been able to keep premature lambs alive for weeks within an artificial womb. Containing a mixture based on the constituents of amniotic fluid, the foetus is bathed in gallons of this fluid as it is pumped through the bag. Replacing the placenta is the equivalent of a heart–lung machine that connects to the umbilical cord and provides oxygenated blood to keep the foetus alive whilst taking the deoxygenated blood away. Although still in the very earliest stages of development, the system appears to be functional and viable and offers the hope that one day even the tiniest of lives could be saved by buying a few extra weeks in the 'womb'.

It's an extraordinary piece of research that really does appear as if it should come from the pages of science fiction, not the science journal *Nature*. This is the cutting edge of our fight for survival, a fight that we all make as individuals every day but that we also make as a civilisation in our collective efforts to push the boundaries of medical research. The story of our survival really is a remarkable one.

# FUTURE

Unpicking the secrets of the human body has previously required an almost singular focus on the question 'why is it like that?' And the answer has usually been found in an understanding of our evolutionary past. Our strategies for learning, growth and survival are grounded in the constraints and opportunities afforded by the environments in which our ancestors lived. It is tempting to believe that we are at an entirely new moment in history: it would seem that, at least for many people in the global north, we have escaped many of the threats that shaped our bodies and killed our ancestors (and more importantly, killed the people who didn't become our ancestors so that their genes didn't survive). In other words, we seem to have reached the end of human evolution. In fact, worse than that, the adaptations we evolved to help us survive now cause us problems.

Childbirth, which was the moment when so many women bled to death for thousands of years, now kills so few people in some developed countries that the maternal mortality rates are close to zero. This event, far more than being attacked by sabre-toothed tigers (a species you might regard as historical *bêtes noires* of evolutionary biology: they applied almost no selection pressure whatsoever in comparison to disease, climate and other humans), shaped your coagulation system.

The predisposition to clotting that gives many people – particularly women – clotting problems like deep vein thromboses is provided by genes that would have kept our ancestors alive, allowing them to clot before they bled to death during childbirth. Our genes could not anticipate the dangers of long-haul flight.

Similarly, the vast portion of our immune system that previously dealt with parasitic infections – tapeworms, malaria and others – is now unoccupied as the majority of people in the UK are entirely free from parasitic illness. It is possible that overactivity of immune systems that have little else to do accounts for a number of illnesses like asthma and eczema that are far less common in populations with a higher parasite burden.

But it seems to me extremely unlikely that we have reached the end of human evolution. There are three ingredients needed for evolution: natural selection, mutation and random change. The second two continue to occur, possibly at higher rates than at other times in history due to a range of new carcinogens and people reproducing at older ages. Natural selection is the evolutionary ingredient that is typically said to have ended. Are there any constraints on survival and fecundity in the modern world that are genetic? Sure, some cultures tend to have more children than others but to a large extent people who want to have children

Human mastery and perversion of natural forces is changing the direction and destination of natural selection. It is the evolutionary ingredient typically said to have ended. Are there any constraints on survival and fecundity in the modern world that are genetic?

in a developed western country can have as many as they like. They might make their lives difficult, expensive or complicated, they may have to endure the rigours of IVF, but if they are inclined to they can usually manage. The vast majority of those children will go on to reach reproductive age and have the same options themselves. This might seem insensitive to the people who have desperately tried to reproduce and not succeeded but it is likely that they were constrained by bad luck and that their story is not part of a bigger picture which is weeding certain traits out of the gene pool. We are no longer losing people who cannot survive a harsh winter, or a water shortage, or who can't hunt enough food. Pretty much everyone gets by (to reproduce at least). You might feel there are people out there with a genetic advantage over you. I feel your pain. I am going bald, getting fat and every day I meet people who are taller, stronger, better looking and smarter than me. Worse still they have leveraged all these qualities to get rich! But in fact, my chances of passing on my genes are about the same as theirs and my son will not be disadvantaged when he tries to have kids compared with their children. He just might not be able to afford a nanny. So why then hasn't human evolution ended?

The modern era in which no one seems to have much of a genetic advantage in surviving and reproducing over anyone else is very recent. The therapeutic revolution is barely one hundred years old, that's just a handful of human generations. The processes of evolution do occasionally force themselves on the earth's life forms very abruptly with a big 'die off'. This uncheerful possibility might take the form of global warming, a meteor strike or a nuclear war. The few people that survived such events would likely have some genetic differences to the current population. Those differences might be related to geography, with a large die off of people from regions near the coast or near the impact site and this would eliminate some genes from the gene pool. Or the differences might be psychological: perhaps in the Thunderdome only the most aggressive people will survive. Or maybe mutations which allowed us to thrive on less food or with less sunlight might become more widespread. Eventually we'll end up with gills or radiation-proof skin or something but the smaller changes are easier to foresee.

However, I'm not saying that we've just hit the pause button on evolution and that it will all start again when 'the big one hits'. We are constantly evolving: the arms race with everything competing with us for that fixed amount of ecological capital on this planet does not pause for a single second. We're all in the same taxi fighting for the best seat. And the taxi isn't getting bigger so we're all having to get sharper elbows. Diseases have shaped much of human evolution, in fact viruses probably drive more selection and gene alteration than any other single environmental factor. We may not face much parasitic illness any more but we are ripe

**Opposite page:** The 'natural' environment of modern humans is very different to that of our ancestors – even those who lived 200 years ago, let alone 200,000.

Chickens confined to prevent them from being exposed to the H5N1 virus. The natural hosts of the H5N1 virus are wild birds, which show few symptoms. However, if transmitted, domestic birds will suffer a 90–100 per cent mortality rate. In addition, humans that come into contact with infected birds can become infected themselves, the first such case occurring in Hong Kong in 1997.

for exploitation from diseases that have not until recently had much of a chance. Those diseases will be viruses and they are likely to shape much of human life in the centuries to come.

Guess how many chickens are raised for eating every year in America (these are called 'broiler chickens'). I know you can just skip ahead and see the number but try to have a guess. The population of the US is currently just over 300 million. OK if you've done the maths here's the answer: they produce almost 9 billion chickens for eating every year. This is such a vast quantity of chickens it is, at least for me, simply impossible to imagine. But it is quite important in terms of human health and disease to consider these numbers of birds very carefully.

In 2015 there was an outbreak of avian flu in the United States. It made few headlines as this strain of flu did not directly affect human health, but the containment measure implemented by the government to control the outbreak required the cull and disposal of approximately 48 million birds. One single farm in Iowa was compelled to kill its entire flock: 5.5 million birds housed in several barns. Many farms had to kill over a million birds. It is not easy to find a method of killing that many birds rapidly (the farms usually achieved this in under 48 hours using foam pumped into the barns to smother them). And it is even harder to dispose of the bodies. The consequences of this were immediately devastating for the farmers and for egg and poultry prices, not to mention for the birds themselves (although I suppose their ultimate fate was always sealed one way or another) but the outbreak was successfully contained. What was the concern? In the first instance to limit the spread of a disease that could devastate the US poultry industry but there is another even more unpleasant scenario to consider. I know what you're thinking: viral mutation. Avian flu becomes a deadly strain of human flu. It is an unpleasant possibility: if you imagine how many opportunities a virus has to mutate as it reproduces within the bodies of potentially 9 billion chickens it might have a good crack at becoming dangerous to humans. But it is unlikely to get that far on its own. The epidemic won't be allowed to get into nearly that many bodies.

There is however another viral phenomenon which is far more concerning: reassortment. This occurs when two related viruses that are infecting the same cell exchange genetic material. Flu viruses seem particularly prone to doing this. Each flu virus has eight different segments, much like genes. If a bird is infected with avian and human flu at the same time those viruses can mix up their segments and produce a new kind of flu. In 2009 this happened not just with bird and human influenza viruses but with pig influenza as well. The resulting strain of flu, popularly known as 'swine flu', has killed over 18,000 people according to the World Health Organization, who admit that the real number is likely far higher. The 'antigenic shift' means that our immune systems are far less capable of recognising and responding to the new virus and the new flu was far more deadly for people under 65 than more typical seasonal flu strains. Chris has had swine flu and he survived. It didn't force a selection pressure enough to alter the big genetic picture of human beings on planet earth. But the changing scale of agriculture, speed of travel, density of people living in cities and the increasing proximity of human populations to livestock all increase the risk that a more severe epidemic will occur. How exactly our bodies will respond to the new viruses is a mystery but it seems possible that emerging disease could exert a severe selection pressure on certain human genes.

Leigh van Valen, the evolutionary biologist who I mentioned in the introduction, developed the 'Law of Extinction' describing the way that every species has an equal likelihood of becoming extinct and that extinction rates are constant over time. It is reasonable to believe that human beings have not escaped this fact of biology and that over the next dozens and hundreds of generations significant natural selection will continue to occur. Now, rather than looking into the evolutionary past for the secrets of the human body we are forced to look into the future to see what secret vulnerabilities or strengths we may contain.

There is another strong candidate for ongoing natural selection, or at least a selection pressure on our population driving differential survival of certain genes in favour of others: genetic engineering and embryo selection. Our technologies to screen embryos for particular traits are only improving as we gain information about the role of genes in the creation of a person. At the moment we can screen for certain diseases and for sex. Once future parents are able to choose hair colour, height, intelligence or athletic ability we can imagine a significant distortion in the population frequency of genes that are able to reproduce themselves. It is hard to imagine that there will not be a market for this service. There is already a market for sex selection. The consequences for a society that prioritises a small number of genetic characteristics and homogenises itself would be catastrophic: if we reduce diversity we render ourselves more vulnerable to threats (and dating will become far more boring). But if I am honest I wonder if, had I been given the choice, I would have opted to give my son an extra twenty points of IQ, or at least spare him my male pattern baldness. The looming possibility of genetic engineering will distort these considerations further. The same technology that will eventually rid us of cystic fibrosis and Duchenne muscular dystrophy will surely give us the options to alter children in ways that we currently cannot imagine. I think there is good reason to be nervous about all of this but that isn't the point. The point is we have not reached the end of evolution and so the human body will always be generating new secrets to investigate.

So the future holds the prospect of new threats (reassorted viruses and more) and new human bodies (as natural selection continues to operate). But there is another central issue of what secrets the human body may contain in the future: the boundaries of our bodies are no longer where we assumed them to be. 'Our' viruses seem no longer to be ours. Instead our viruses are composites; blurred entities relevant to many species but created in some ways by our behaviour. You could reasonably say that this distinction is a bit academic; that viruses have reassorted themselves for a long time and just because they now have more opportunities to do it doesn't represent a paradigm shift. That may be so, though I would argue that the scale and pace of these changes means we are living in a world that virologically does not resemble even our fairly recent past. But there are other more profound areas in which we are having to consider the boundaries of our bodies. The first is the microbiome: the ecosystem of microorganisms that live in and on us. When I was at medical school less than two decades ago this was barely discussed. We considered our bodies to be governed by a genome containing approximately 20,000 genes. Today we can reasonably expand the number of genes that exert some control over significant aspects of our bodies (including our weight, appe-

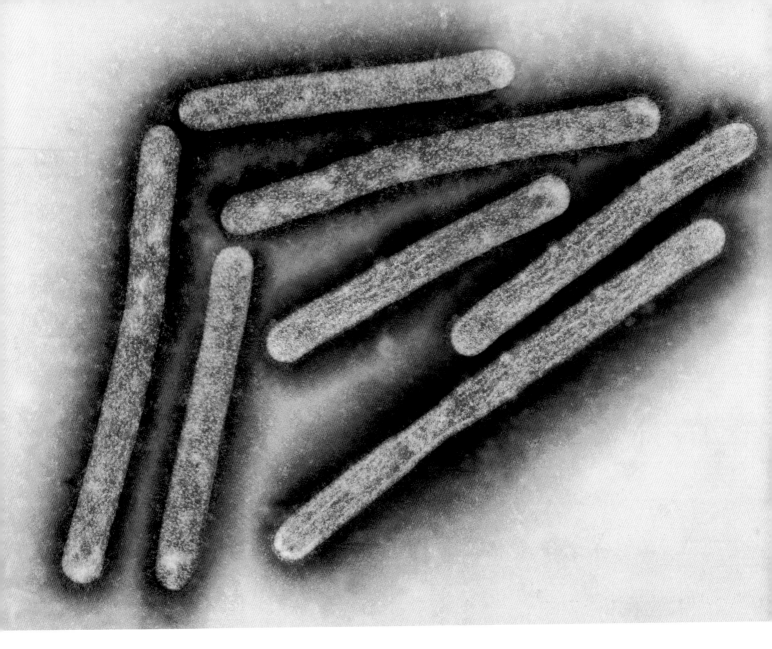

tite, digestive system, mood and resistance to infection) to about 5 million. That is the number of bacterial genes that exist in our biome. Without them we could not live and even small changes to them seem to be detrimental to our health. The secrets that lie within this collection of organisms are vast and extraordinarily difficult to explore. You can swab the skin or the gut and grow the resulting cultures on petri dishes but that tells you little about the ratios *in vivo* or their relationships in context. It is fortunate that there is a gene that is common to all biome bacteria, that has variations between bacterial species, and that is not found in plants, animals, viruses or fungi. By sequencing this gene we can build up a picture of the numbers of species found in, say, the human gut, at any one time. Although at the moment if you were to have your gut biome characterised there would be many species that the scientists would not be able to identify. As we saw in the 'Grow' chapter the single bacteria that can digest the main sugar in breast milk is not

Coloured transmission electron micrograph of H5N1 avian influenza virus. This subtype of bird flu can spread to humans through close contact with infected birds or their faeces. Concerns are that this subtype might mutate and turn into a human influenza subtype that could cause a worldwide flu epidemic.

present in the guts of many humans. There are many other secret components of our bodies out there and characterising the biome is a task equivalent to characterising a rainforest: a vast complex network of relationships and competitions. And like so many of our rainforests we are in danger of destroying it before we fully understand it. Increasing use of antibiotics and shifts in our diets are significantly altering our microbiomes with reduced diversity and proliferation of species that are possibly pathological.

The boundaries of our bodies are socially constructed. What a 'human body' is will shift with its uses and roles in different areas of life. There is a regular conference at the Google Campus in Silicon Valley called SciFoo. It is a chaotic amazing multidisciplinary conference in which there is no fixed schedule but simply a huge wall chart of time slots and lecture rooms into which you can scribble a title and your name, and hope that people show up to listen to you talk. There are Nobel prize winners, NASA scientists, gamers, engineers and a ton of other amazing people talking about every topic under the sun. I went to every talk I could and what was noticeable was that there was one topic generating more heat than any other: man–machine interfaces. You could hardly get into the rooms.

Coloured scanning electron micrograph of a concave electrode array on a MEMS (microelectromechanical systems) neural interface device. This device is designed to connect to the brain to allow direct electrical communication with the brain's neurons.

Man–machine interfaces are things like touch screens and keyboards but increasingly there is an interest in putting the interface within the body. You could tell this was the hot topic at SciFoo because the founders of Google, two of the smartest, most ambitious, richest guys on the planet showed up to an already packed room and pounded the lecturers with questions. Not just general questions but extremely specific questions about 'bits'. As in 'how many bits per second can you get out?' and 'how many bits per second can you get in?' They were asking about the rate of exchange of information that it is possible to achieve between a microchip and a human brain. Enough to make a robot walk? Enough to move a cursor?

It turns out it is very hard to move information into or out of the brain in ways other than the ways it has evolved to communicate. But it is possible.

We already have neuroprosthetics in the form of cochlear implants and artificial retinas. They are crude but improving rapidly. And we have invasive brain–computer interfaces. A team at Emory University in Atlanta, led by Philip Kennedy and Roy Bakay, installed a brain implant in their patient, Johnny Ray in 1998. Ray was 'locked-in' meaning he was unable to move any part of his body after a brainstem stroke. He eventually learnt to control a computer cursor. Animal experiments and the growth of 'neural chips', circuitry that incorporates living nerve tissue, suggest that our connections to machines are likely to grow deeper and faster quite rapidly.

It seems impossible to imagine that brains could communicate with brains or computers as fast as computers talk to one another, but the speed of this interface is likely to be one of the areas of huge scientific concentration over the next few decades. One of the striking things is that the current interfaces require a great deal of learning. Just as children need to learn to accurately move their limbs, the adults learning to control machines with their brain – or as they experience it, their mind – must invest many hours of practice. Suggesting that childhood might be a better time to start wiring ourselves up. Now this may fill you with a nameless horror – and I relate to that – but the boundary between the machine and the human body is already extremely thin. My mobile phone can take my pulse, count my steps and is activated by my thumbprint. The apps on it recognise my face. I never turn it off and it is almost never out of my sight. It is an extension of my mind and body in a way that no other device I have ever owned or encountered comes close. Perhaps Roger Federer has a similar relationship with his tennis racket, but I bet he spends less time holding it than most people spend holding their cell phones.

What are the implications for the human body once it is connected to machinery? Our measures of strength and intellect, of our ability to resist disease become less relevant. Instead the next wave of human evolution may take place as the ones most able to reproduce are the ones most able (or willing) to connect themselves to sex robots or artificially intelligent computers. They will be the ones whose genes predispose the cortex of their brains to accept the fine wires of the bio-compatible microchip. Perhaps in the more distant future the boundaries of the body will be entirely redefined as our brains become so integrated with computers that the existing genetic material is irrelevant. We will all be able to exist for ever as our souls are uploaded to the cloud. This seems a long way off but the recordings of much of our lives, the data on our sleep and on our movement patterns is already streaming into warehouses full of servers in the Nevada desert; we are building DNA computers, mechanical hearts and most depressingly (or excitingly) we are manufacturing sex robots. With regard to these last devices, where you sit depends on where you stand, as they say.

The point is that the human body will no longer be viewed as a self-contained internally consistent bag of cells each of which contains our unique genome. Instead we will increasingly see it as a complex network of interdependent organisms (the not popular but technical term is 'holobiont') with deep connections to non-organic (or partially organic but non-living) artificially intelligent networks and mechanical systems. Personally I think this sounds pretty dismal. I'd like a self-driving car, but I'm prepared to forgo it to avoid actually living in the matrix. But it is already happening and you can bet that I have failed to forecast even one per cent of the strangeness of the future our bodies are about to be living in. Medicine has always seemed like a safe career in terms of job security but in fact the only people that AI won't be able to replace will be children's TV presenters. Everyone else will go the way of the factory workers and the horse-drawn carriage.

Regardless of whether you're excited or terrified by these possibilities (I think probabilities) the drive to achieve them will force scientists to think about the secrets of the human body in entirely new ways.

One of the things that struck me at medical school was that almost every new discovery made things more complicated. You might imagine that scientific discoveries simplify things. That they kind of tidy everything up. E=mc² seems pretty neat and compact. But in the biological sciences the decoding of the genome revealed layers of complexity we could hardly have imagined. The investigations into the microbiome seem to have done the same thing. Wrestling with this complexity is a job for supercomputers much of the time and there is one burgeoning field that demonstrates just how complex the data can become: proteomics.

Your genetic code works by making proteins. Proteins are the molecules that build your body: they are the builders and the engineers that haul everything else together and make it interact properly. They are muscular fibres, enzymes and scaffolding. They can respond to changes in the external environment: in other words they can take instruction, they can do work.

The process of making proteins seems straightforward: your genes get read by an enzyme called DNA transcriptase and turned into lengths of RNA. These lengths of RNA get translated by a ribosome which understands each group of three letters codes for a particular amino acid. So far so good: proteins are long chains of amino acids. If we are able to understand their behaviour we can understand the mechanics of life itself. But here's the annoying bit. We don't really understand proteins at all. All kinds of things happen to these ribbons of amino acids once they are made. They have many different chemical groups added to them often in lengthy sequences of post-production alteration: ubiquitination, glycosylation, acetylation, methylation, oxidation and phosphorylation. All of these will change the way in which they chemically interact with the world around them. The starting point for understanding a protein though is not an accounting of all these, though that is important, but rather its structure. Basically, asking the question 'what shape is it?'

The structure of a protein tells us a lot about its function. A protein that breaks down glucose will have a structure that can recognise glucose and binds to it so that the chemically relevant amino acids can do their job. That chain of hundreds or potentially thousands of amino acids is not just bundled up like a ball of string. It takes a very exact conformation, its 'native conformation' that may change as it does its job. Every link in the chain is somehow 'sticky' to other specific links and so there is a perfect arrangement of the chain that is the most stable and satisfactory, though there are many millions of possible arrangements. You have to sift through all the possibilities to figure out which is the most stable.

This would seem to be at first glance a complete nightmare. And at second glance that idea seems to be confirmed: when you ask a supercomputer to calculate the most stable arrangement of a protein it turns out to take days. It is expensive to use supercomputers and even if scientists can raise the funds to buy several days' worth of a computer's time, or compete with their colleagues for limited available hours on their lab's computer, the results are frequently not that accurate. Artificial intelligence seems to hold some promise in this area and the possibility of making new proteins or improving existing proteins seems to lurk just around the corner with all the benefits that could bring for the management of diseases, but it is still one of the most expensive and complicated parts of biological science.

The history of proteomics is recent (the term was coined in 1997, a year after Chris and I began medical school) and impressive but there is one part of this

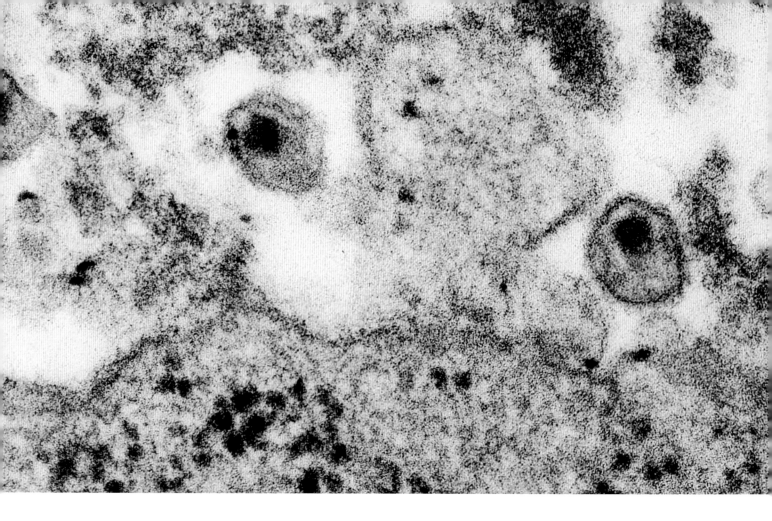

endeavour which is particularly charming and should offer some hope to anyone fearing 'the rise of the machines' and the redundancy of anyone who is not a children's television presenter.

In 2008 David Baker, a leading protein scientist at the University of Washington, was using a computer program called Rosetta to reveal protein structures and needed more computing power. He decided to harness the power of some of the millions of computers that sit idle in people's homes all day long and made the program available for download. If you downloaded Rosetta@home it would display the dimensional protein structures it was working on as screensavers, incredibly beautiful floating, rotating, rainbow-coloured structures that showed these delicate molecules fundamental to life.

These mesmerising screensavers were typically shown in the home of interested molecular biologists (most 'regular people' have not downloaded a protein folding program), and they caught the attention of the observers who began to do an extremely human thing: look for patterns. And some of them found patterns and solutions to the problems that the computers were working on. They could see more stable configurations of the ribbons of protein than the ones the computers were generating. There was a problem though: they were unable to interact with the program, they just had to watch the computer botch it up. So they began

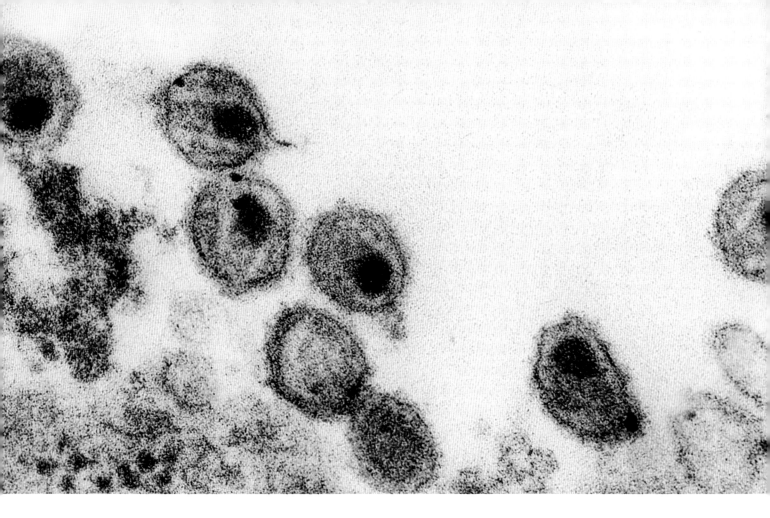

contacting David Baker who did an extraordinary thing. He enlisted the help of Zoran Popović, a computer game designer at his university, and they built a computer game to allow members of the public to work on protein structure. This seems like work if you do it for a living but it's like playing Tetris or Candy Crush if you do it for fun. The game is called Foldit. You can still go online and play. It's not as slick as a regular computer game website but it will teach anyone – very little biology knowledge needed – how to fold proteins. And then start you working on some of the most important scientific questions of our time. This is a cute story and this method of work has a name, distributed thinking. It harnesses the human brain's unique ability to manage complex problems in three dimensions.

In 2011 the Foldit team published a paper in *Nature Structural & Molecular Biology* entitled 'Crystal structure of a monomeric retroviral protease solved by protein folding game players'. The players had managed to solve a problem that had eluded scientists and their supercomputers for over a decade: describing the structure of an enzyme critical for reproduction of HIV. It took them three weeks to reveal the enzyme's structure and identify targets for drugs to neutralise it. Perhaps more remarkable even than this (and they have many other comparable achievements now) is what it tells us about the secrets of the human brain: in one of the user discussions on the Foldit website someone asks about the background of the more successful gamers. One answer stood out to me: 'I'm a former computer programmer, now housewife.' Proteomics is one of the most expensive, complex, important scientific problems in the world and one of the leaders in the field is a housewife. I asked one of the Foldit team about her: 'Oh yes, she lives in the UK in the Midlands.' It's hard to imagine a better kept human secret than a housewife living in the Midlands who is better than a supercomputer at unpicking the mysteries of life.

The future will create new tasks and new opportunities for our human bodies so there will never be a moment when we have revealed all its secrets. The possibility of integrating computers, or of finding new ways of exploiting our existing computer power, the possibility of introducing new engineered organisms into our biome, the new challenges presented by climate change or emerging viruses: all of these things will reveal secret answers to questions that could not even have been posed by previous generations.

Anyone attempting to guess what the future holds is destined to fail but it seems reasonable to believe that along with new discoveries new barriers will emerge. The technologies we now have to alter and interact with the human body exceed our abilities as a society to agree on ethical norms. The profit motive significantly distorts research and means that certain questions are privileged over others. And the more that is known about our bodies – the more secrets that are revealed – the more vulnerable we become to exploitation. My concerns centre around the value of information about ourselves to private companies and to the state and there is no doubt that as genetic sequencing becomes cheaper (and anyone can now get their genome sequenced for a couple of hundred pounds) we will be faced with the possibility of knowing secrets about ourselves that we might be unable to handle. Increasingly accurate calculations of life expectancy, and of maximum physical and mental capability or aspects of our psychology, are clearly double-edged swords. Perhaps to some they will reveal untapped potential and boost confidence. They will allow efficient selection of workers, increased profitability and optimal 'progress'. But to others they will be an iron cage: a ceiling imposed on their ambitions from birth. The list of traits that such a society might value and select for seems seductive in the specific and the short term – intelligence, attractiveness – but fatal in the long term. Human society could become almost monocultural. Auden talks about the crowds along the pavements as 'fields of harvest wheat' but I don't think he'd be keen to see this vision of uniformity realised too literally.

There will be no end to the secrets: our bodies will continue to create and try to keep new ones. The business of uncovering them is so thrilling and the projects to understand ourselves are now so vast that a wider view of this science can find a god-like quality in all this knowledge. We can set our genes to music; we find beauty in the folding of a protein; we find our truest motives in the possibility of tinkering with the destiny of embryos. As we look deeper into our cells and the genes that guide them we find Blake's 'Auguries of Innocence' made manifest:

> To see a World in a Grain of Sand
> And a Heaven in a Wild Flower
> Hold Infinity in the palm of your hand
> And Eternity in an hour.

The cells in the palm of your hand are descended from a lineage that goes back not just to the first humans but ultimately to the first strands of life, and that entire history – almost infinitely into the past and into the future – is written in the choices of genes that have endured. We can see the world in cells far smaller than grains of sand.

But we should end not with secrets but with mysteries. The tremendous scope of these scientific endeavours, that, like Ulysses, 'pursue knowledge like a sinking star beyond the utmost bound of human thought' do have limits. We may feel that with proteomics and genetic engineering and the huge datasets that we can now acquire and interrogate that we can reach out and touch the face of god.

But the grand ambition and design of the modern scientific project can also seem like a kind of parochial essentialism. There are things that science cannot uncover: biochemistry, cell biology and physiology are not the appropriate research methods with which to ask questions about meaning. The compromises that allow our bodies to function will never be resolved. We will always be a little too disgusted to have fun, and not quite squeamish enough to stay safe. We will always have immune systems that have to find a halfway point between keeping us free from infection and eating ourselves. We will always have a mechanism of fear that prevents us doing some of the risky important stuff but also fails to keep us completely out of harm's way. Being human is to live in a society within an ecosystem and that must always be a compromise. Hamlet's dithering about being and not being is, of course, correct. It's a contradictory mess. And so for answers about meaning we must turn elsewhere: the poets and painters, philosophers, historians, anthropologists. These people will always have an essential role in framing the questions of our lived experience. Shakespeare (or Hamlet) doesn't answer the most famous question posed in literature. But he does hold our hand and show us how perfectly balanced the contradictions are.

# INDEX

# ACKNOWLEDGEMENTS

We'd like to sincerely thank all of the production team for the endless hard work that has gone into making the television series that accompanies this book. With Nigel Paterson and Giselle Corbett at the helm, the series has been lead by two of the very best people in the business. It's never easy marrying cutting-edge research with world-class television production but they have once again delivered an outstanding series.

We'd particularly like to thank Martin Johnson, Penny Palmer and Matt Dyson for producing such beautiful and thought-provoking films and for managing the stresses and strains of the production with such good humour. They were supported by a hugely talented team who have grappled week after week with the demands of production, so a very big thank you to Charlotte Lathane, Fleur Bone, Ross Kirby, Francesca Bassett, Joanna Fulcher, Laura Kennedy, Rebecca Hicki. It is impossible to know every aspect of the human body and Xand and I have had no shame in leaning on the expertise, research and learning of this fantastic team of directors and producers. We count ourselves unbelievably lucky to work with the in-house science team at the BBC.

We'd also like to thank Ged Murphy, Darren Jonusus and Lee Sutton for their craft and creativity in the edit. And Patrick Acum, Kevin White, Simon de Glanville, Tom Hayward, Paul Kirsop, Stuart Thompson, Freddie Claire, Chris Youle Grayling, James Kenning, for their beautiful work on location. None of this would have been possible without the creativity of Nicola Cook, Adriana Timco and our brilliant development team who shaped and nurtured the idea from its very beginning and made sure the idea grew into the project it is today. We'd also like to thank Laura Davey, the Production Executive of BBC Science, Marie O'Donnell, our talent manager and Vicky Edgar, who holds everything together.

The team at HarperCollins have once again been nothing short of brilliant, performing what can only be described as a publishing magic trick by seeming to produce this beautiful book out of thin air when faced with (huge apologies!) extremely late delivery from the authors. We'd like to thank Julia Koppitz, Tom Cabot, Philippa Hudson and, of course, the ever-patient and wise Myles Archibald (who quite literally had to spend even more time in the dentist's waiting room).

Neither the book nor the series would have been possible without the research and time of scientists around the world. Harald Ott and his whole lab bent over backwards to accommodate us at MGH and showed us things that needed to be seen to be believed. Sarah Gilpin in particular gave me a crash course in stem cell biology. Bruce German and Mark Underwood at UC Davis took us from the basic bench chemistry through to the huge clinical benefits of HMOs in a single day. Tracey Shafizadeh helped filming and ensured I had some *B. Infantis* for when Lyra started breast feeding. Dr Lauren Sherman and Dr Mirella Dapretto at UCLA stayed late to help us film their remarkable MRI study on social media. Professor Mike Tipton, Dr Heather Massey and the whole Extreme Environment Lab at the University of Portsmouth gave time and expertise with their usual generosity to show the remarkable homeostatic capabilities of the human body. Dr Frederick Ahs and his team at the Psychology Division of Karolinska Institute gave my amygdala a thorough workout and reframed neuroscience for us. Professor Val Curtis showed us through lived experience how powerful disgust is. We haven't been able to eat pigs' head since then! Buy her book: *Don't Look, Don't Touch, Don't Eat*. Professor Dan Davis and his team at the University of Manchester showed us the most astounding immunology images we'd ever seen – and were a perfect inspiration for the write-up stage of Chris's own PhD. Professor Greg Towers at UCL was a source of time and inspiration for many of the ideas in this book. Thanks for being a great supervisor. Thanks to Tahu and Stefani for making Chris learn to juggle. I'll beat you one day. Thanks to Lew Hollander for some of the best advice about ageing we've ever heard and for letting us film his spectacular efforts. Chris and Xand would like to thank Miranda Chadwick and Kate Mander – man's best agents and friends. Thanks to Mum, Dad, Dinah, Carolina and J for their support. And to Lyra for putting up with Chris doing this through the first few weeks of your life. Andrew would more than anything like to thank Anna … for being … well, just Anna, the most beautiful of human beings. And for also taking on so much (that's a hefty mental load!!) whilst I disappeared into my laptop.

# A FIREFLY BOOK

Published by Firefly Books Ltd. 2017

First printing

**Publisher Cataloging-in-Publication Data (U.S.)**

Library of Congress Cataloging-in-Publication Data is available

**Library and Archives Canada Cataloguing in Publication**

A CIP record for this title is available from Library and
Archives Canada

Published in the United States by
Firefly Books (U.S.) Inc.
P.O. Box 1338, Ellicott Station
Buffalo, New York 14205

Published in Canada by
Firefly Books Ltd.
50 Staples Avenue, Unit 1
Richmond Hill, Ontario L4B 0A7

Printed and bound in Great Britain by Bell and Bain Ltd, Glasgow

First published by William Collins,
An imprint of
HarperCollins Publishers
1 London Bridge Street
London SE1 9GF
By arrangement with the BBC

Publishing Director: Myles Archibald
Senior Editor: Julia Koppitz
Cover Design: Julian Humphries
Production by Chris Wright
Edited, designed and illustrated by Tom Cabot/ketchup

## PICTURE CREDITS

Page 2: Bernhard Jank/Ott Lab, Center For Regenerative Medicine; page 4: BBC; pages 6–7: Steve Gschmeissner/Science Photo Library; page 8: Mehau Kulyk/Science Photo Library; pages 10-11: Mauritshuis Museum, The Hague; page 13: Public Domain; page 14: Sebastian Kaulitzki/Science Photo Library; page 15: Contour/Getty Images; page 16: Science Photo Library; page 19: Laguna Design/Science Photo Library; pages 20–1: Steve Gschmeissner/Science Photo Library; page 23: Public Domain; page 25: Claude Nuridsany & Marie Perennou/Science Photo Library; page 26: Matthew Daniels/Wellcome Images; page 30: Contour/Getty Images; pages 32–3: Dr G. Moscoso/Science Photo Library; page 34: Chris And Xand van Tulleken; page 36: Dr Yorgos Nikas/Science Photo Library; page 38: Kit van Tulleken; page 39: Chris van Tulleken; pages 40-1: Shutterstock; page 40: Bruce German; page 43: Eye of Science/Science Photo Library; page 45: Tom Cabot/Ketchup/Eureka; page 47: Chris van Tulleken; page 48: Burger/Phanie/Science Photo Library; page 52: Nick Obank/Barcroft Media/Getty Images; page 55: Www.Telegraph.Co.Uk/Tom Cabot; page 58: Microscape/Science Photo Library; page 60: United Artist/Getty Images; page 61: Graphicaartis/ Getty Images; pages 62–3: Zephyr/Science Photo Library; page 64: Wellcome Library, London/Wellcome Images; page 65: Chris van Tulleken; page 69: Erich Auerbach/Getty Images; page 72: Arran Lewis/Wellcome Images; page 73: Tom Cabot/Santiago Ramón Y Cajal; page 75: Nancy Kedersha/Science Photo Library; page 76: Kul Bhatia/Science Photo Library; page 78: Dr P. Marazzi/Science Photo Library; page 81: Garo/Phanie/Science Photo Library; pages 82–3: W. W. Schultz/British Medical Journal/Science Photo Library; page 85: Microscape/Science Photo Library; page 89: Lew Hollander; page 93: Rob Young/Wellcome Images; page 94: Jan Halaska/ Science Photo Library; page 100: Thierry Berrod, Mona Lisa Production/Science Photo Library; page 103: Nancy Kedersha/Science Photo Library; page 104: Bernhard Jank/Ott Lab, Center For Regenerative Medicine; page 107: Bernhard Jank/BBC; page 109: Contour/Getty Images; pages 110-11: www.Humanconnectomeproject.org/gallery; page 112: Mehau Kulyk/Science Photo Library; page 114: Chris van Tulleken; page 115: Rebecca Saxe; page 119: Tom Cabot; page 121: Chris and Xand van Tulleken; page 123: BBC/Andrew Cohen; page 126: Claus Lunau/Science Photo Library; pages 128–9: Deb Roy & Rupal Patel/HSP; page 132: Krishna Vukoti; page 133: Tom Cabot/Ketchup; page 134: Tom Cabot/Ketchup; page 136: Ludovic Collin/Wellcome Images; page 141: Tom Cabot/Ketchup; page 142: Jacapo Annese/Courtesy of UC San Diego; page 144: Sovereign/Ism/Science Photo Library; page 146: Courtesy of Deborah Weaving; page 150: CORBIS/Corbis Via Getty Images; page 151: Nasa/Carla Cioffi; page 153: Jordan Mansfield/Getty Images for LTA; page 154: Sciepro/Science Photo Library; page 155: Thomas Deerinck, Ncmir/Science Photo Library; page 156: D. Phillips/Science Photo Library; page 158: King's College London Archives/Science Photo Library; page 158: Science Source/Science Photo Library; page 161: Nigel Pavitt/Getty Images; page 162: Denny Levitt; page 165: Chris Packham; page 167: Mint Images/Science Photo Library; page 168: Paul D Stewart/Science Photo Library; page 170: Bettmann/Getty Images; page 172: A. Dowsett, Health Protection Agency/Science Photo Library; page 174: Nlm/Science Source/Science Photo Library; page 176: Burger/Phanie Science Photo Library; page 177: Cc Studio/Science Photo Library; page 178: BBC/Andrew Cohen; page 179: Bernhard Limberger/LOOK-Foto/Getty Images; page 180: Tom Beard/Chris van Tulleken; page 182: BBC/Andrew Cohen; page 184: BBC/Andrew Cohen; page 186: Freddieclaire/BBC/Andrew Cohen; page 188: Alfred Pasieka/Science Photo Library; page 190: Alex Hyde/Science Photo Library; page 191: Science Source/Science Photo Library; page 192: John Watson/Johns Hopkins University; page 193: Wellcome Dept. of Cognitive Neurology/ Science Photo Library; page 194: Roger Harris/Science Photo Library; page 196: Anatomical Travelogue/Science Photo Library; page 197: David Parker/Science Photo Library; page 199: Thierry Berrod, Mona Lisa Production/ Science Photo Library; page 200: Wellcome Dept. of Cognitive Neurology/Science Photo Library; page 203: John W Banagan/Getty Images; page 206: Jorge Bernal/AFP/Getty Images; page 208: Claude Nuridsany & Marie Perennou/Science Photo Library; page 211: Nibsc/Science Photo Library; page 213: Royal Free Hospital/Chris van Tulleken; page 214: National Institutes of Health/NIAID/Science Photo Library; page 216: Olivier Blaise/Getty Images; page 218: Chris and Xand van Tulleken; page 219: Chris And Xand van Tulleken; page 221: Eye of Science/Science Photo Library; page 222: Mezida B. Saeed and Daniel M. Davis; page 226: Quim Llenas/Getty Images; page 227: Science Photo Library; page 229: Biophoto Associates/Science Photo Library; pages230–1: Chris And Xand van Tulleken; page 237: Felix Pharand-Deschenes, Globaia/Science Photo Library; page 238: Thierry Berrod, Mona Lisa Production/ Science Photo Library; page 240: Philippe Benoist/Eurelios/Science Photo Library; page 243: Dennis Kunkel Microscopy/Science Photo Library; page 244: David Scharf/Science Photo Library; 248–9: Callista Images/Cultura/Science Photo Library.